# Hot Isostatic Pressing
## HIP'17

12th International Conference on Hot Isostatic Pressing (HIP'17), Sydney, Australia, 5-8 December, 2017.

## Pranesh Dayal and Gerry Triani

Australian Nuclear Science and Technology Organisation (ANSTO)

Peer review statement

All papers published in this volume of "Materials Research Proceedings" have been peer reviewed. The process of peer review was initiated and overseen by the above proceedings editors. All reviews were conducted by expert referees in accordance to Materials Research Forum LLC high standards.

Copyright © 2019 by authors

Published under License by **Materials Research Forum LLC**
Millersville, PA 17551, USA

Published as part of the proceedings series
**Materials Research Proceedings**
Volume 10 (2019)

ISSN 2474-3941 (Print)
ISSN 2474-395X (Online)

ISBN 978-1-64490-002-4 (Print)
ISBN 978-1-64490-003-1 (eBook)

This book contains information obtained from authentic and highly regarded sources. Reasonable efforts have been made to publish reliable data and information, but the author and publisher cannot assume responsibility for the validity of all materials or the consequences of their use. The authors and publishers have attempted to trace the copyright holders of all material reproduced in this publication and apologize to copyright holders if permission to publish in this form has not been obtained. If any copyright material has not been acknowledged please write and let us know so we may rectify in any future reprint.

Distributed worldwide by

**Materials Research Forum LLC**
105 Springdale Lane
Millersville, PA 17551
USA
http://www.mrforum.com

Manufactured in the United State of America
10 9 8 7 6 5 4 3 2 1

# Table of Contents

# Preface

This volume is a record of the papers presented at the 12th International Conference on Hot Isostatic Pressing (HIP) held in Sydney, Australia during 5 - 8 December, 2017. The HIP'17 conference was hosted by the Australian Nuclear Science and Technology Organisation (ANSTO) hosting delegates from 15 countries. The focus of this three-day technical symposium was to discuss emerging trends, developments and innovations in the field of HIP technology.

Since the invention of HIP in the mid-1950s, there has been a consistent evolution of knowledge covering areas in materials and process development including advances in equipment capability. Today this technology is applied in many industry sectors including advanced materials manufacturing, automotive, aerospace, oil and gas, power generation, medical and nuclear. The papers within this volume represent the advancement across many of these industry sectors.

I would like to personally thank all of the delegates for attending and contributing to this successful conference and in particular to our sponsors, our local and international committees for their tireless work and time given to ensuring the success of this conference series.

Finally, I would also like to thank the Materials Research Forum LLC (MRF) for publishing these proceedings. This book is dedicated to the authors of this volume who continue to advance the field of HIP Technology.

Gerry Triani
HIP 17 Conference Chair

# Committees

**Conference Chair**
- Gerry Triani

**Program Committee**
- Dr. Dorji Chavara
- Prof. Christoph Broeckmann
- Dr Victor Samarov

**Local Organising Committee**
- Dr. Dorji Chavara
- Dr. Pranesh Dayal
- Kelly Cubbin
- Mitchell Smith

**IHC Committee**
- Martin Bjustrom, David Bowles, Christoph Broeckmann, Michael Conaway, Brandon Creason, Andres Eklund, Genrickh s. Garibov, Wolfram Graf, Michael Hamentgen, Ken Hirota, Chen Hongxia,  Stephen Mashl, Itura Masuoka, Cliff Orcutt, Dennis Poor, , Victor Samarov, Serge Sella, Gerry Triani, Xinhau Wu and Yamamoto Yasuhiro

**Emeritus IHC members**
- Takao Fujikawa, John Hebeisen, Kozo Ishizaki, Gerard Raisson

Materials Research Forum LLC
doi: http://dx.doi.org/10.21741/9781644900031-1

# HIP for AM - Optimized Material Properties by HIP

Magnus Ahlfors[1,a] *, Fouzi Bahbou [2,b] and Anders Eklund[3,c]

[1]Quintus Technologies LLC, Lewis center, OH, USA

[2]Arcam AB, Mölndal, Sweden

[3]Quintus Technologies AB, Västerås, Sweden

[a]magnus.ahlfors@quintusteam.com, [b]fouzi.bahbou@arcam.com,
[c]anders.eklund@quintusteam.com

**Keywords:** Hot Isostatic Pressing (HIP), Additive Manufacturing, Titanium

**Abstract.** An investigation of HIP parameters for EBM Ti-6Al-4V has been performed by Arcam AB and Quintus Technologies AB with the aim to maximize the strength of the HIP:ed material. A lower HIP temperature of 800 °C and a higher pressure of 200 MPa gives the highest strength and is also enough to eliminate all internal defects. By printing material with intentionally induced porosity combined with an optimized HIP cycle the highest strength can be obtained.

## Introduction

Hot isostatic pressing (HIP) is widely used today to eliminate internal defects in metallic materials produced by powder bed fusion. The internal defects are mostly lack-of-fusion defects generated during the printing process and entrapped gas porosity coming from the powder particles. These defects act like stress concentrations and crack initiation points in the material, which decreases the material properties. By eliminating these defects within the material, the ductility and especially the fatigue properties are improved [1-5]. Figure 1 shows a cross section of an EBM Ti-6Al-4V material before and after HIP where the typical effect of HIP:ing in terms of defect elimination can be seen. In Figure 2, typical fatigue data of as printed and HIP:ed material of EBM Ti-6Al-4V is shown and it is evident that the HIP process gives much improved fatigue properties compared to as-printed material. This data is generated by Arcam.

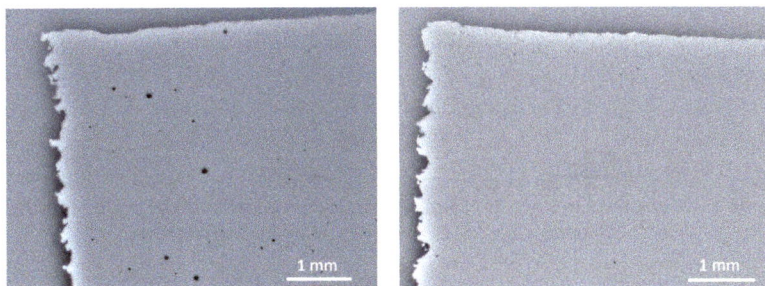

*Figure 1 - Micrographs of EBM Ti-6Al-4V before HIP to the left and after HIP to the right*

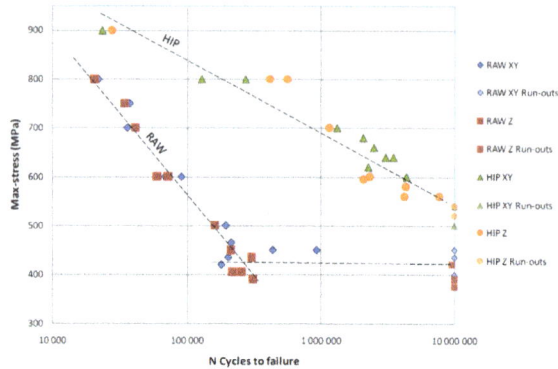

*Figure 2 - Fatigue data for EBM Ti-6Al-4V, courtesy of Arcam*

The solidification rates in the EBM process are in the order of $10^3 - 10^5$ K/s which is very high [6]. The extremely fast solidification generates a very fine microstructure which gives the material a high strength. Any conventional heat treatment at an elevated temperature for a significantly long time, like HIP:ing, will coarsen the microstructure due thermodynamic driving forces. This coarsening of the microstructure will decrease the strength of the material, which is not preferable. The development within EBM printing equipment over the last years has made the as-printed microstructures even finer, which makes this challenge even more significant for the modern EBM machines. In Figure 3 a) and b) the microstructure of as printed material compared to HIP:ed EBM Ti-6Al-4V is shown. The coarsening of the microstructure after HIP is evident. Figure 3 a) and c) shows the difference between the microstructures produced by an older Arcam s12 machine compared to a newer Arcam Q10 machine.

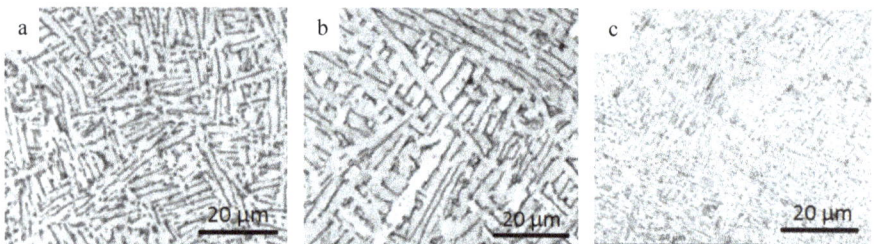

*Figure 3 - Microstructures of EBM Ti-6Al-4V a) As-printed with Arcam s12 b) After HIP (920°C, 1000bar, 2h) with Arcam s12 c) As-printed with Arcam Q10*

For Ti-6Al-4V produced by selective laser melting (SLM), the same coarsening of the microstructure and thus decrease of strength has been seen. As reported by Leuders [1] the tensile strength of SLM Ti-6Al-4V is decreased by any kind of elevated temperature process including HIP:ing as shown in Figure 4.

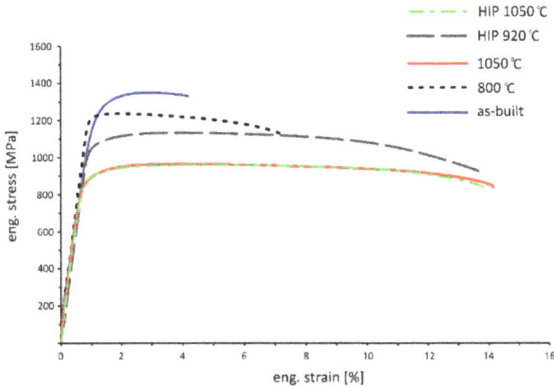

*Figure 4 - Tensile test data of SLM Ti-6Al-4V in different conditions [1]*

The ASTM standard for additive manufactured Ti-6Al-4V by powder bed fusion, F2924-14, specifies a HIP temperature between 895 and 955 °C, a minimum pressure of 100 MPa and a minimum hold time of 2 hours. The widely used HIP parameters for AM Ti-6Al-4V is 920 °C, 100 MPa and 2 hours hold time, thus within the standard specification. However, these HIP parameters were developed for cast Ti-6Al-4V before today's commercial AM processes existed and it is not obvious that these HIP parameters are optimal for AM material.

With this background, a study of HIP parameters for EBM Ti-6Al-4V was made by Arcam and Quintus Technologies with the purpose to evaluate if other HIP parameters could be used to eliminate all defects but have a lower influence on the fine as-printed microstructure. The approach of this study is to evaluate lower HIP temperatures.

**Experimental**

The HIP parameter study consisted of 6 different HIP treatments according to Table 1 together with as-printed material as reference. The test specimens for the study were 15 mm diameter rods printed with standard EBM printing parameters and powder and 10 test specimens for each condition were used for the evaluation.

*Table 1 - The 6 different HIP treatments used in the study*

| Variant | Temperature [°C] | Pressure [MPa] | Hold time [hours] | Cooling rate [K/min] |
|---------|------------------|----------------|-------------------|----------------------|
| 1 | 920 | 100 | 2 | ~ 30 |
| 2 | 920 | 100 | 2 | ~ 1500 |
| 3 | 880 | 100 | 2 | ~ 30 |
| 4 | 840 | 100 | 2 | ~ 30 |
| 5 | 800 | 100 | 2 | ~ 30 |
| 6 | 800 | 200 | 2 | ~ 30 |

The results of the different HIP variants were evaluated by density, internal defect analysis and tensile data in xy and z-direction. The density was measured by helium pycnometry and with Archimedes principle, here called water intrusion. The internal defects in the material were evaluated by optical microscopy and the tensile data was attained with standard tensile testing.

As an extension of the HIP parameter study, material printed with intentional internal porosity was also HIP:ed and analyzed. The intentional induced defects were generated by printing with a larger line off-set up to 0.4 mm instead of the standard 0.2 mm. In Figure 5 the macrostructure of material printed with the standard 0.2 mm line off-set and 0.4 mm line off-set is shown and a large difference in number of defects can be observed.

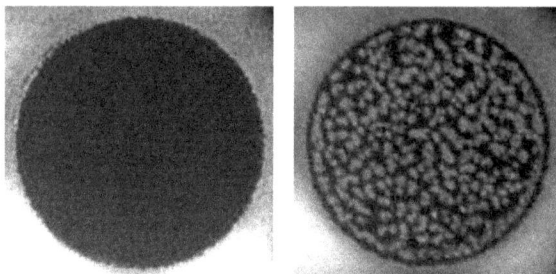

*Figure 5 - Defects analyzed with LayerQam. To the left with 0.2 mm line off-set and 0.4 mm to the right.*

**Results**

In Figure 6 the results from the density measurements are presented. As can be seen, all HIP variants gives ~ 100 % density. The material printed with standard parameters also shows ~ 100 % density in the as-printed state. There is still porosity in the material but a relatively small amount as shown in Figure 1 so that no influence is visible in the density measurements. For the material printed with larger line off-set the density is significantly lower than the standard printed material with up to 8 % porosity for the material printed with 0.4 mm line off-set. However, these samples are also fully densified to ~ 100 % density with HIP despite the low starting density. The contour of the porous printed specimens was built with standard printing parameters to achieve a gas tight surface for the specimens. A gas tight outer surface of the part to be HIP:ed is required to ensure full densification by the HIP process without any encapsulation of the part.

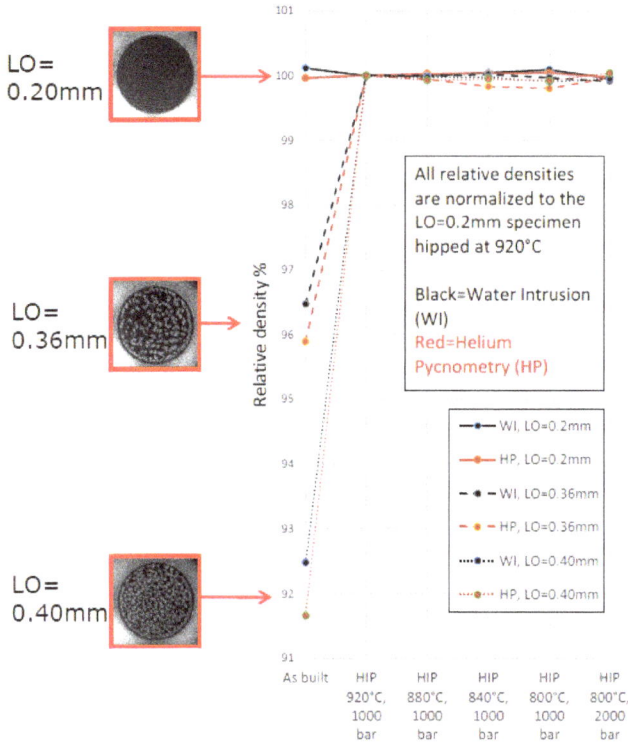

*Figure 6 – Density measurements results*

The results for the evaluation of defects with optical microscopy showed that HIP variants 1, 2, 3 and 6 gave full defect elimination. HIP variants number 4 and 5 with 840 and 800 °C and 100 MPa showed some small remaining defects which were not eliminated by HIP. The density measurements in Figure 6 also tend to show the same result.

The tensile test results for the material printed with standard parameters are shown in Figure 7. The results show that there is a significant increase in strength when going towards lower HIP temperatures. The HIP cycles at 800 °C gives the best strength and the ductility is not influenced significantly by the different HIP variants.

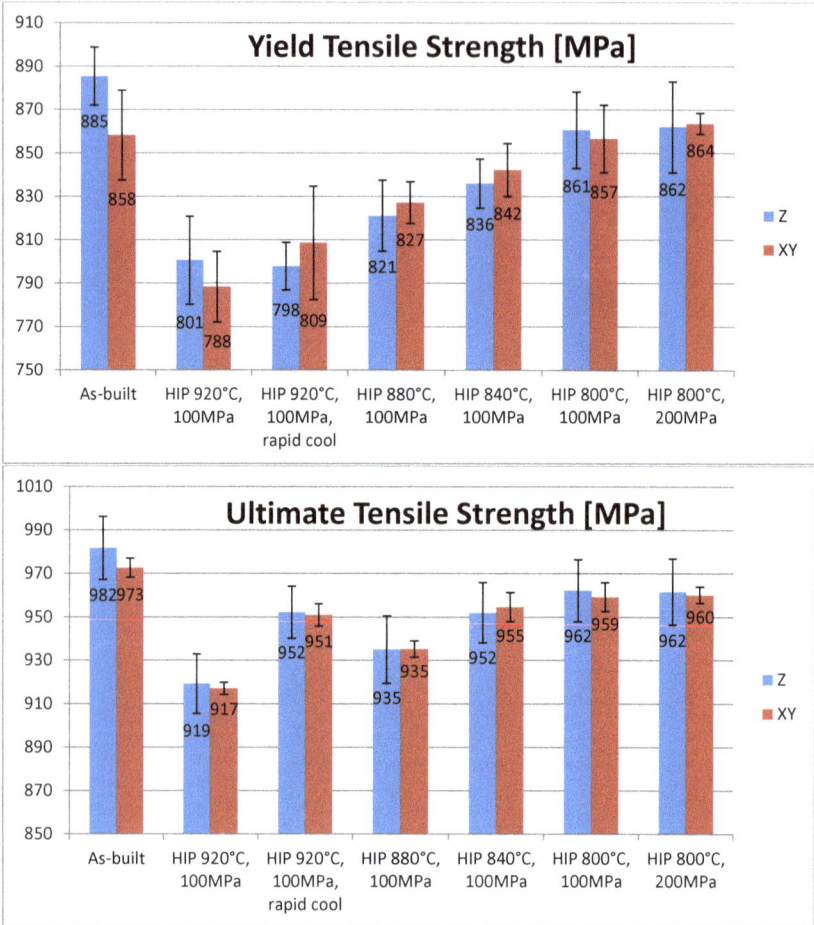

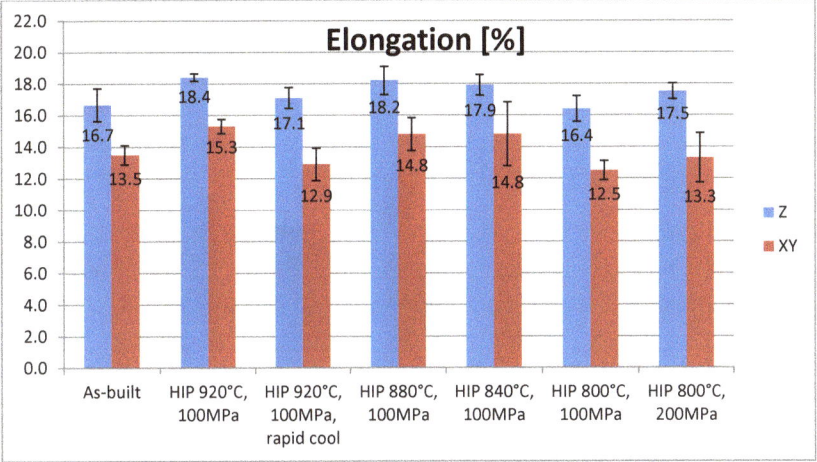

*Figure 7 - Tensile data for standard printed material*

The tensile test results for the material printed with intentional defects are shown in Figure 8 together with the data from the standard printed material. The first interesting observation that can be made is that the strength of the porous printed material is higher than the standard printed material even though it contains ~ 8% porosity. However, the ductility of the porous printed material is significantly reduced by the ~ 8% porosity. After HIP, the strength values are reduced but always higher for the porous printed material compared to the standard printed material. For the porous printed material HIP:ed with optimized parameters at 800 °C the strength is higher than the as-printed material printed with standard parameters. In addition, the very low ductility in the as-printed porous material is fully recovered by the HIP process making a material with higher strength than the as-printed material and the same ductility.

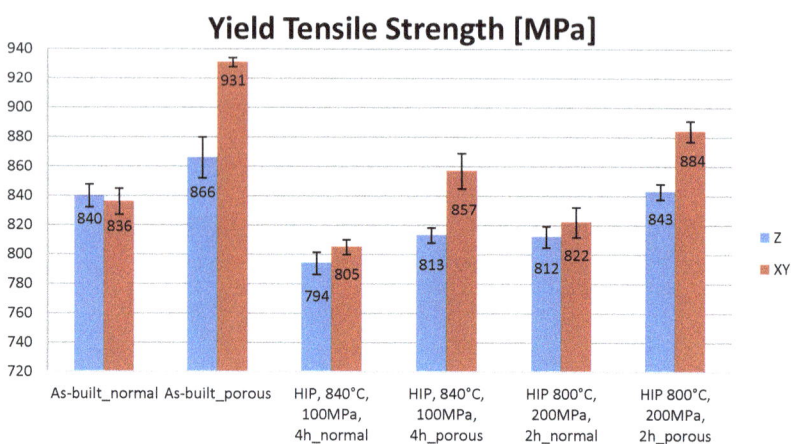

**Yield Tensile Strength [MPa]**

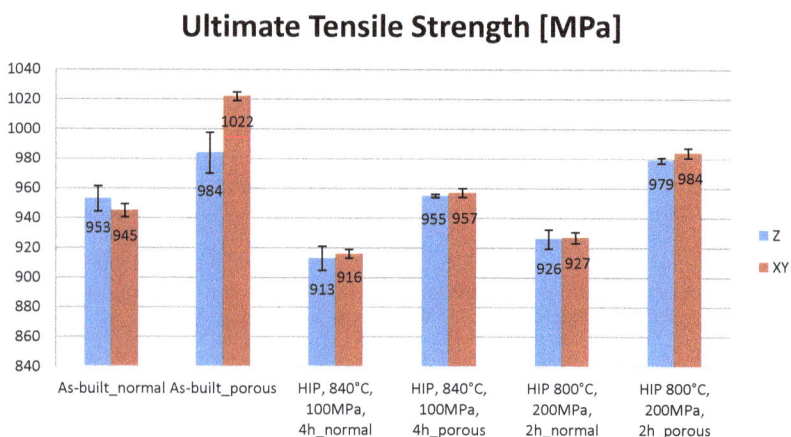

**Ultimate Tensile Strength [MPa]**

Materials Research Forum LLC
doi: http://dx.doi.org/10.21741/9781644900031-1

## Elongation [%]

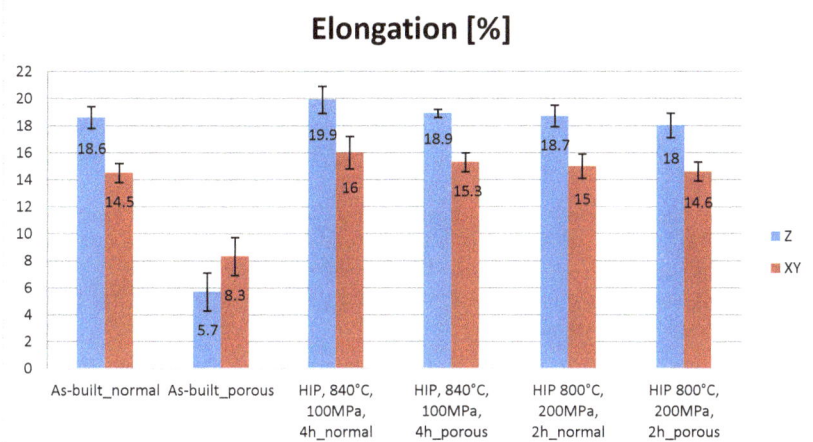

*Figure 8 - Tensile data for standard and porous printed material*

An interesting question is, why the material printed with ~ 8% porosity shows higher tensile strength than a virtually fully dense material printed with the standard parameters? An explanation for that could be the increased line off-set of 0.4 mm which will generate a smaller melt pool in the printing process and so the layers are less re-melted. This reduced re-melting of the material results in an even finer as-printed microstructure, which gives a higher strength. The smaller melt pool and less re-melting also reduces the loss of aluminum due to evaporation during the printing process which, which is positive in terms of strength.

**Conclusions**
The conclusion from this study is that it is possible to HIP EBM manufactured Ti-6Al-4V with a HIP temperature as low as 800 °C and still get full density in the material if the HIP pressure is increased from 100 to 200 MPa with an unchanged hold time of 2 hours. These HIP parameters will give a significantly higher strength compared to the standard HIP parameters of 920 °C, 100 MPa and 2 hours due to reduced coarsening of the fine microstructure.

It can also be concluded that printing with a larger line-offset induces more porosity in the material during the printing process but also gives higher strength due to less energy input into the material. The result is that optimized HIP parameters for AM in combination optimized printing parameters for HIP will lead to a material which is stronger than the as-printed standard material but shows same ductility level. It can be assumed that the HIP:ed material will also have the already well-known significant increase in fatigue properties compared to the as-printed material.

The final conclusion from this study is that HIP, heat treatments and other processes should be investigated for AM material in order to achieve optimal properties, and these post treatments cannot just be copied from e. g. the casting industry because additive manufacturing is something different than the conventional manufacturing processes. Another important point is if HIP:ing is used, the printing process can be adjusted for HIP to achieve optimum material properties. In case of a subsequent HIP process, there is no need to achieve 99.9 % theoretical density in the as-printed part since the densification result after HIP:ing will be the same even with 92 % as-printed density.

**References**

[1] S. Leuders, On the mechanical performance of structures manufactured by Selective Laser Melting: Damage initiation and propagation, University of Paderborn, Germany, as presented at AMPM2014, MPIF, USA

[2] N. Hrabe et al., Fatigue properties of a titanium alloy (Ti–6Al–4V) fabricated via electron beam melting (EBM): Effects of internal defects and residual stress, International Journal of Fatigue Volume 94, Jan 2017. https://doi.org/10.1016/j.ijfatigue.2016.04.022

[3] A. Kirchner et al., Mechanical Properties of Ti-6Al-4V Additively Manufactured by Electron Beam Melting, EuroPM2015, Reims, France

[4] J. J Lewandowski M. Seifi, Metal Additive Manufacturing: A Review of Mechanical Properties, Annual Review of Materials Research, July 2016

[5] J. Haan et al, Effect of Subsequent Hot Isostatic Pressing on Mechanical Properties of ASTM F75 Alloy Produced by Selective Laser Melting, EuroPM2014, Salzburg, Austria

[6] S. Al-Bermani, M. Blackmore, W. Zhang, I. Todd, The origin of microstructural diversity, texture, and mechanical properties in electron beam melted Ti-6Al-4V, Metallurgical and Materials Transactions A 41 (2010) 3422–3434. https://doi.org/10.1007/s11661-010-0397-x

Materials Research Forum LLC
doi: http://dx.doi.org/10.21741/9781644900031-2

# Hot Isostatic Pressing of Radioactive Nuclear Waste: The Calcine at INL

Dr. Anders Eklund[1,a*], Dr. Regis A. Matzie[2,b]

[1]Quintus Technologies AB, Quintusvägen 2, SE-72166 Västerås, Sweden

[2]RAMatzie Nuclear Technology Consulting, LLC, 15 Winhart Drive, Granby, CT 06035, USA

[a]anders.eklund@quintusteam.com, [b]regismatzie@gmail.com

**Keywords:** Heat Treatment, Hot Isostatic Pressing (HIP), Radioactive Waste, Calcine, Vitrification, Immobilizing, Radioactive Isotopes, Leak-Before-Burst, Pre-Stressed Wire-Wound, Safety

**Abstract.** Hot Isostatic Pressing (HIPing) is a technology that has been around for 60+ years. By using high temperature and high gas pressure, dry metal and ceramic powders can be consolidated and a volume decrease can be achieved. This paper presents the simulations of using the HIPing process at the Idaho National Laboratory to treat the calcine radioactive waste. Once loaded into collapsible canisters, the HIPing would be used to treat the waste before final disposal. The resulting volume reduction was shown to be 20-70% and the cost ratio vs vitrification is 1:1.74.

## Introduction

### Hot Isostatic Press (HIP)

Hot Isostatic Press has fundamentally two different designs when it comes to contain the high pressurized gas, typically Argon. The two methods are called mono-lithic, sometimes referred to as mono-block, and pre-stressed wire-wound technology. An example of the wire-wound pressure vessel can be seen in Fig. 1.

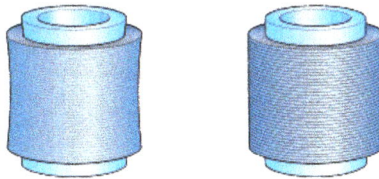

*Fig. 1. Pre-stressed wire-wound vessel design. Without pressure applied (left) and with pressure applied (right).*

The pre-stressed wire-wound HIP will always experience compressive stresses both on the inside and outside of the pressure vessel and the yoke frames during all phases of the HIPing process. This is the safest design and is approved by ASME per ASME Boiler and Pressure vessel code, Section VIII, Division 3. This failure mode is described as "Leak-before-burst". This means that if the pressure vessel cracks, the gas under high pressure will dissipate through the wire-wound package without any damages to the surrounding equipment and structures. For example, see Fig. 2.

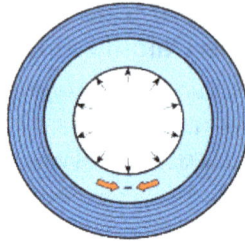

*Fig. 2. Pre-stressed wire-wound vessel design showing the compressive stresses of a material fault even under high pressure.*

The combination of elevated temperatures, 300-2500 °C, and high gas pressures, 50 – 300 MPa, consolidates dry metal and ceramic powders by mechanical deformation, creep and diffusion, and heal internal voids, i.e. metal castings, to substantially improve the strength of any materials. The temperature depends on the material to be HIPed, e.g., Aluminum has lower melt temperature (650 °C) than steel (1550 °C). An example of the effect of HIPing of voids in a material can be seen in Fig. 3.

*Fig. 3. Cross section of artificial pore before HIPing (left) and after HIPing (right).*

The HIP cycle itself is strongly dependent, as mentioned before, on the parameters temperature and pressure. But, also time is of the essence in most applications since the material densification is depended on creep and diffusion which are time dependent mechanisms. Conventionally a HIP furnace was cooled naturally which took a lot of time of the total cycle time. A large HIP with a heavy load could take up to 12 hours to cool down before it was possible to open and start a new HIP cycle. Much efforts have been done the last decades to optimize and minimize the total cycle time. The introduction of a forced convection cooling technology called Uniform Rapid Cooling (URC®) greatly decreased the cycle time and allows the HIP operators to optimize the cycle to be most suitable for them and their materials. Cooling rates of 100 °C/min or more can easily be achieved. This increases the productivity of the HIP unit significantly since the material throughput is increased for the same HIP size. An example of a HIP cycle without and with URC can be seen in Fig. 4.

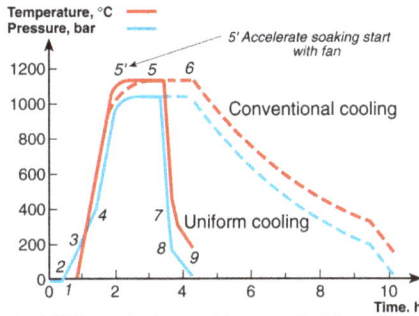

*Fig. 4. Typical HIP cycle times without and with uniform rapid cooling.*

The HIP system itself consists of the wire-wound yoke frames and a thin-walled cylinder, which can be considered as the back-bone of the pressure vessel since they take up the forces coming from the compressed gas. See Fig. 5.

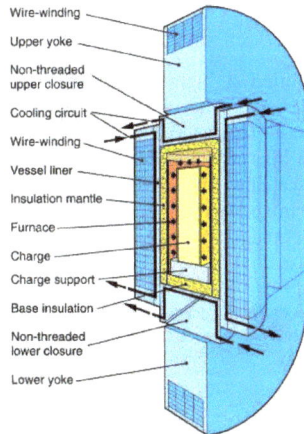

*Fig. 5. Pre-stressed Wire-wound HIP machine.*

The furnace, which is the heart of the HIP machine, has an elaborate design to ensure good insulation, temperature accuracy, rapid cooling and reliable and safe requirements. HIP furnaces can be supplied in Steel, Molybdenum or Graphite, depending on the operating temperature. See Fig. 6.

Hot Isostatic Pressing – HIP'17                                      Materials Research Forum LLC
Materials Research Proceedings **10** (2019) 11-17          doi: http://dx.doi.org/10.21741/9781644900031-2

The charge, which can be canisters of powder or cast/forged parts, is placed on an insulated support structure. The gas flows freely around the charge for utmost temperature accuracy. The best temperature accuracy is achieved with a multi-zone convection furnace. See Fig. 6.

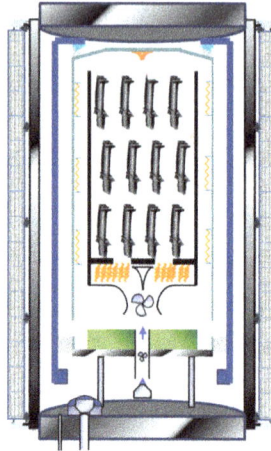

*Fig. 6. Closer Look on the Furnace with multiple heating zones.*

**The Radioactive Waste**

The Idaho Calcine is by definition a high-level waste. It is the first raffinate from the reprocessing of the spent nuclear fuel. The un-treated calcine is also classified as hazardous waste as it exhibits hazardous waste characteristics for toxicity of metals. Today, roughly 4400 $m^3$ of granular solid is stored in bins at the Idaho National Laboratory (INL) site. It has been shown by several studies that a calcine can be HIPed directly, if the particle size and material composition is right, or it can be mixed with an additive and then HIPed in a collapsible canister [1,2], see also Fig. 7. In both cases the result is a glassy ceramic waste form that can be packaged and disposed of in a final repository.

*Fig. 7. HIPed treated zirconia calcine canister inside the MFC HFEF hot cell at INL.*

### Description

**Hot Isostatic Pressing of Calcine**

To make the test mixture, additives were combined with surrogate calcine to make a glass-ceramic final waste form in the HIP. The mixture was placed into a HIP can for bake out and sealed for HIP testing. Quintus Technologies, LLC assisted in development of the bake out and can sealing approach in their lab-unit in Columbus, OH as well as in the HIP-unit at INL supplied by Quintus Technologies, see Fig. 8.

*Fig. 8. Quintus HIP unit installed in the HFEF hot-cell at the INL*

After HIP testing, the canisters were sectioned so that samples could be collected from the HIP material. The temperature was varied to test the volumetric reduction for different pressures and temperatures. See Fig. 9 [3].

*Fig. 9. Vertical cuts of all cans showing various volumetric reductions.*

## Discussions

### Results of HIPing

From the trials mentioned above, some important conclusions could be made. First of all, that HIPing works to consolidate a ceramic like calcine. Also, that it is a versatile method for many different applications and can be used for many different possible treatment routes. See Table 1.

Secondly, that HIPing proves to be a very cost-effective method with the lowest life-cycle cost for the consolidation of the calcine, since the main cost is for storage volume and HIP shows the lowest storage volumes of all compared technologies. Today's generally established route of vitrification is shown to cost about 75% more than HIPing. See Table 2.

*TABLE 1. HIP Technology Advantages over Vitrification [4].*

| Consolidation: | HIP | Vitrification (JHM) |
|---|---|---|
| Matrix: | Glass-ceramic | Borosilicate glass |
| Waste Loading: | 60-90% | 20-35% |
| Durability: | 10-100 x EA-glass | 10 x EA-glass |
| Final Volume: (relative to untreated calcine | 20-70% reduction (treat=low, non-treat=high) | Min. 100% increase |
| Temperature: | 1050-1200 °C | 1150 °C |
| Pressure: | 35-50 (100) MPa | Atmosphere |
| Off-gas/By-Product waste: | Minimal | Medium-High |
| Flexibility: | | |
| - Calcine: | Treat or super-compact | Treat only |
| - Future mission: | Diverse/Flexible | Limited/Less flexible. |

*TABLE 2. Cost Comparison: HIP vs Vitrification [4].*

| Parameters | HIP without RCRA Treatment | | HIP with RCRA Treatment | | Direct Disposal | | Vitrification with Separations | | Vitrification without Separations | |
|---|---|---|---|---|---|---|---|---|---|---|
| Canisters | 2 900 | 3 300 | 3 700 | 4 600 | 6 700 | 7 300 | 750 | 2 200 | 11 100 | 13 300 |
| Total Life Cycle Cost (MUSD) | 5 503 | 6 228 | 6 052 | 7 119 | 7 661 | 8 408 | 9 556 | 12 769 | 11 054 | 13 074 |
| Cost ratio | 1.00 | 1.13 | 1.10 | 1.29 | 1.39 | 1.53 | 1.74 | 2.32 | 2.01 | 2.38 |

### Conclusions

Several advantages can be seen by using pre-stressed wire-wound HIP for consolidating the calcine at INL site. HIP is a proven technology since 60+ years with world-wide safe operations. Quintus HIP systems are built according to ASME Boiler and Pressure vessel code, Section VIII, Division 3, "Leak-before-burst". Technical advantages are that the waste is isolated from other process equipment, the process is scalable, it has the highest safety, there are no emissions from the consolidating process, it is flexibility to produce a range of waste forms, volume reduction of 20-70%, and it is a batch process for easy operation and diversity and reduces risk for heterogeneous waste feeds. Economically the HIP process shows the lowest life-cycle cost compared with direct disposal and vitrification.

## Acknowledgements

The authors wish to thank CH2M-WG Idaho, LLC for permission to publish the results of this work. This work was supported by U.S. Department of Energy, Idaho Cleanup Project, and Project No. 23582.

## References

[1] M. Burström and R. Tegman, US patent 4642204, Feb.10, 1987.

[2] K. J. Bateman, R. H. Rigg and J. D. Wiest, Proc. ICONE10, Arlington, VA, April 14-18, 2002, 10TH International Conference on Nuclear Engineering, ASME 2002.

[3] D. C. Mecham, B. R. Helm and V. J. Balls, "Calcine Disposition Project Small Can Test Report, RPT-866, U.S. Department of Energy, Assistant Secretary for Environmental Management, Under DOE Idaho Operations Office, Contract DE-AC07-05ID14516

[4] B. D. Preussner et. al., Idaho Cleanup Project, U.S. Department of Energy, Idaho Operations Office, Project No. 23582.

Hot Isostatic Pressing – HIP'17                                  Materials Research Forum LLC
Materials Research Proceedings **10** (2019) 18-23        doi: http://dx.doi.org/10.21741/9781644900031-3

# Hot Isostatic Pressing in the Automotive Industry: Case study of Cast Aluminum Alloys for Rims of Car Wheels

Anders Eklund[1,a] and Magnus Ahlfors[1]

[1]Quintus Technologies AB, Quintusvägen 2, SE-72166 Västerås, Sweden.

[a]anders.eklund@quintusteam.com

**Keywords:** HIP, Aluminum Alloys, Castings, Wheel Rims, Automotive

**Abstract**. Cast aluminum alloys are good candidates for weight saving in many industries, i.e automotive, aerospace, sporting goods and other high-performance application. When HIPing of aluminum alloys is performed the fatigue properties of the casting is greatly improved. That is especially true for rims of car wheels that suffer from high porosity and high scrap rates. After HIP, zero porosity is found and scrap rates drop with 50-90%.

## Introduction

Cast pores are potential crack initiation sites for aluminum alloys and is considered the main influence of poor fatigue properties. Also, the microstructure influences the mechanical behavior of cast alloys, like inclusions, dendrite spacing and grain size. By controlling the cooling rate during solidification of castings, it is possible to control and modify the alloy microstructure and thereby optimizing the mechanical properties. In Fig. 1, AlSi7 cast aluminum alloy is seen before and after HIP. The internal porosity is completely eliminated, but surface connected pores are still visible on the tested samples. The remaining surface pores will disappear after the final painting step.

*Figure 1. Material before HIP (in the middle) and after HIP (top/bottom).*

Hot Isostatic Pressing – HIP'17                    Materials Research Forum LLC
Materials Research Proceedings **10** (2019) 18-23        doi: http://dx.doi.org/10.21741/9781644900031-3

## Production route for rims for car wheels

Two different production routes have been considered for the manufacturing of rims for car wheels. In Fig. 2, the HIP is used to treat the aluminum alloy directly after solidification to eliminate porosity and improve the machined surface quality to lower the scrap rate, see Fig. 3. The best way to control the cooling rate is achieved with a HIP-system equipped with uniform rapid cooling (URC™), see Fig. 4. Uniform rapid cooling was introduced in the 1980's for enhancement of the productivity making it possible to double the production due to shorter cycle times. Another advantage was the better control of the cooling rate, which makes it possible to optimize the pressure-temperature ratio which enables optimization of the material properties.

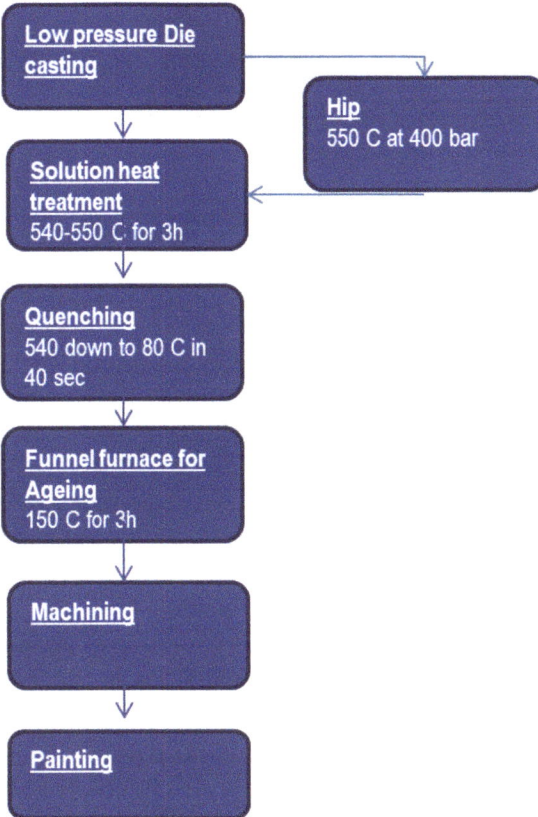

*Figure 2. Possible production route for HIP after casting.*

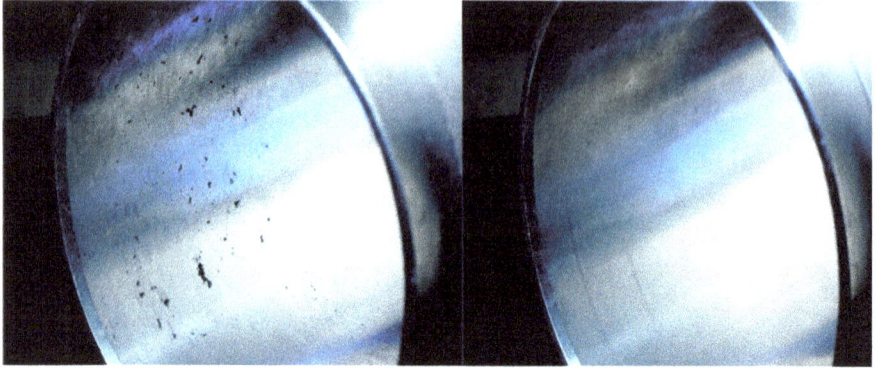

*Figure 3. Improved machined surface quality, before HIP (left), and after HIP (right).*

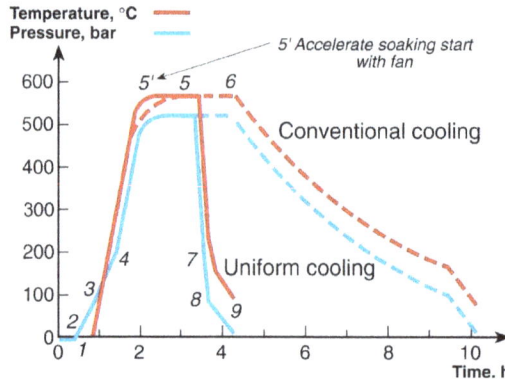

*Figure 4. Typical HIP cycle times without and with rapid cooling.*

The second possible production route can be seen in Fig. 5. Here, the HIP will replace even more process step, i.e. the solution heat treatment (SHT) and the quenching, by utilizing the possibility the combine HIPing and heat treatment, HPHT. The latest developments in HIP technology, Uniform Rapid Quenching (URQ), have made it possible to achieve cooling rates up and over 2000 °C/min. The same quench rates as you experience in oil- and water bath quenching. The advantage with HPHT, is that the HIP pressure is maintained during the complete HIP cycle until the final step when the HIP is opened for the removal of your parts, see Fig. 6.

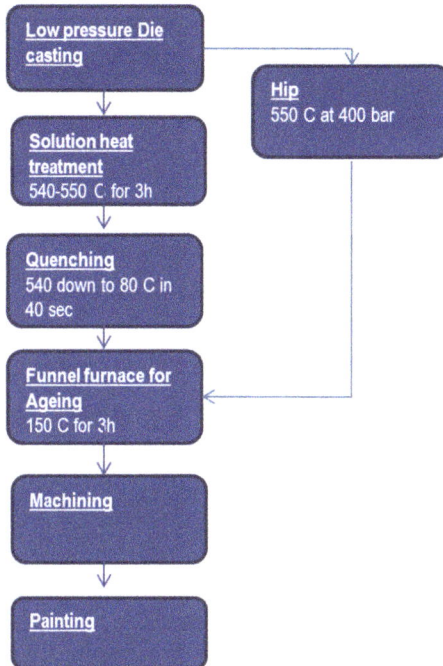

*Figure 5. Possible production route for HIP removing SHT and Quenching.*

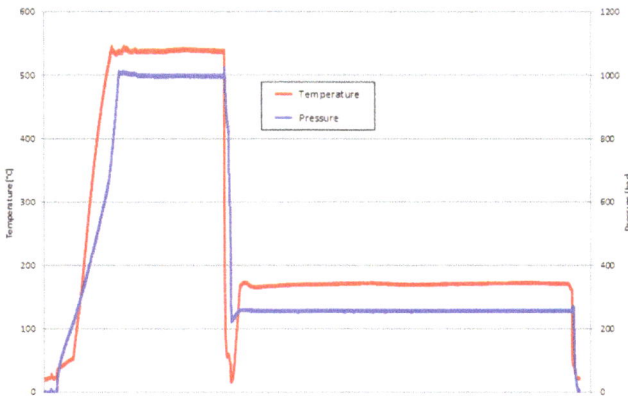

*Figure 6. HIP log curve at 1000 Bar and 538° C with a following T6 heat treatment at 168° C and 260 Bar for 6 hours.*

Many advantages are found when HIP and Heat Treatment is combined in the same cycle. Cost savings due to lowered energy usage and lowered scrap rates due to the elimination of porosity. The material properties are enhanced, especially fatigue life and elongation, but also hardness can be increased by modifying and optimizing the microstructure. The influence of HIPing on the fatigue life can be seen in Fig. 7. The average fatigue life increases from less than 20k cycles to well above 140k cycles for the same stress load, before failure.

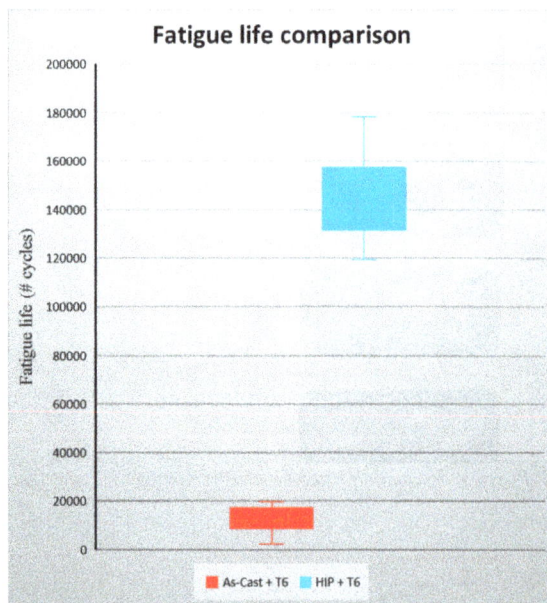

*Figure 7. Fatigue life comparison with and without HIP at the same load stress.*

Another way to describe the improvement of HIPing for rims of car wheels is to study the development of elongation through the production process, see Fig. 8. Every time the Al-alloys is going through a process step, the elongation decreases due to mechanical or thermal impact of the wheel. Especially, the solution heat treatment and ageing make the large drop in elongation, but also machining and painting makes a lowering in elongation. The threshold for approval in this case is 3% elongation.

One clearly see that HIP increases elongation dramatically and thereby adds a cushion so the threshold value is out of reach, and consequently a drastic drop in scrap rate due to being too low in elongation values.

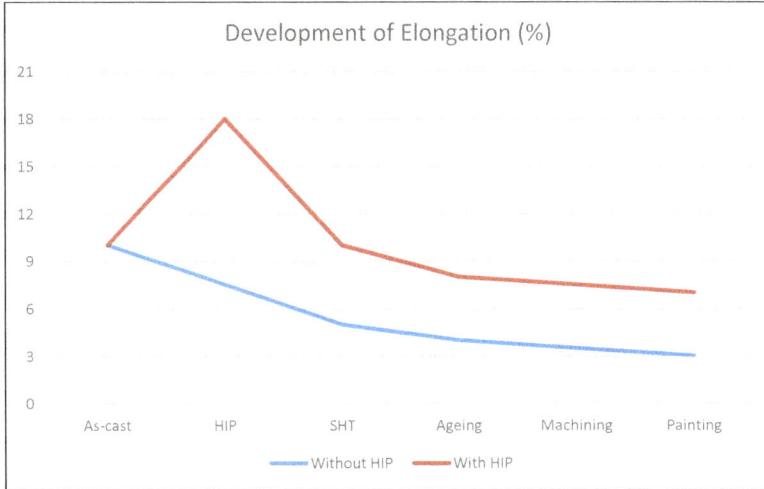

*Figure 8. Development of elongation during manufacturing.*

**Summary and Conclusion**

The automotive industry is always looking for weight reductions so the cars will be lighter and save fuel. Aluminum alloys have been used for making the rims of car wheels lighter for many years. If HIPing and a combined heat treatment step is introduced, the material properties can be optimized, especially fatigue life and elongation. The scrap rates will be reduced, and other cost savings as lesser energy usage as well as reduced material usage can be seen.

Hot Isostatic Pressing – HIP'17                                    Materials Research Forum LLC
Materials Research Proceedings **10** (2019) 24-29          doi: http://dx.doi.org/10.21741/9781644900031-4

# HIP Technology Enables Ceramic Manufacturers to Control Material Properties and Increase Productivity

Dr. Anders Eklund[1,a*], Dr. Magnus Ahlfors[2,b]

[1]Quintus Technologies AB, Quintusvägen 2, SE-72166 Västerås, Sweden

[2]Quintus Technologies, LLC, 8270 Green Meadows Drive North, Lewis Center, OH 43035, USA

[a]anders.eklund@quintusteam.com, [b]magnus.ahlfors@quintusteam.com

**Keywords:** Ceramic, Hot Isostatic Pressing (HIP), Heat Treatment, Zirconia, Dental Implants

**Abstract.** Hot isostatic pressing (HIP) has been known for more than 50 years, and is considered today as being a standard production route for many applications. The HIP process applies high pressure (50-200 MPa) and high temperature (300-2,500°C) to the exterior surface of parts via an inert gas (e.g., argon or nitrogen). The elevated temperature and pressure cause sub-surface voids to be eliminated through a combination of mechanical deformation, plastic flow and diffusion. The challenge is to reach the highest possible theoretical density while maintaining productivity goals.

Uniform rapid cooling is a process by which thin-walled pre-stressed wire-wound HIP units increase productivity up to 70% compared with natural cooling, and increase the density to ~ 100% of theoretical density for many alloys. The added cost to reach this density is around $0.20/kg for a large production HIP system, depending on the material.

## HIP introduction

HIP technology was introduced in the early 1950s and has since gained interest for many applications. HIP is a fabrication process for the densification of castings, consolidation of powder metals (as in in metal injection molding or tool steels and high-speed steels), compaction of ceramics for dental and medical parts, additive manufacturing (3-D printing), and many more applications.

Today, about 50% of all HIP units are used for the consolidation and heat treatment of castings. Typical alloys include Ti-6Al-4V, TiAl, aluminum, stainless steel, nickel-based super alloys, precious metals like gold and platinum, and heavy alloys and refractories like molybdenum and tungsten. More applications will likely come rapidly due to the increased interest in the additive manufacturing of ceramics for aerospace and automotive applications.

The pressure applied in a HIP is generally between 100 and 200 MPa using pure argon gas. However, both lower and higher pressure can be used for some special applications. Other gases like nitrogen and helium are also used, while gases like hydrogen and carbon dioxide are more seldom put into use in production units. Combinations of these gases can also be used. The application determines which gas is used for which purpose, especially since helium is quite expensive compared with argon/nitrogen and hydrogen in incorrect concentrations is very explosive.

The parts to be HIPed are initially heated either at elevated pressure or in vacuum. Introducing the gas early in the process, and while heating, causes it to expand and help to build up the pressure in the HIP furnace more effectively. The material composition and suggested HIP cycle govern the startup procedure. The three main advantages of HIPing include:

*Increase in Density*
- Elimination of internal porosity for defect healing.
- Predictive lifetime
- Lighter and/or low-weight designs

*Improvement of Mechanical Properties* [1, 2, 3, 4]
- Fatigue life increased up to 10 times, depending on the alloy system
- Longer lifetime of HIPed parts
- Decrease in variation of properties
- Increase in ductility and toughness
- Form metallurgical bond between dissimilar materials (diffusion bonding)

*More Efficient Production*
- Decreased scrap/loss
- Less or no non-destructive testing (NDT)
- Freedom to choose casting methods for optimal productivity

## Uniform Rapid Cooling (URC®) and Quenching (URQ®)

The demand from industry has always been to have shorter HIP cycle times and thereby increase the productivity for the HIP unit to achieve better payback for the investment. The HIP pressure vessel itself plays an important role in increasing the cooling rates for the efficient removal of the heat generated in the furnace. The wire-wound technology shown in Fig. 2 provides several advantages to enhance this heat removal.

The heat removal is achieved by the thin-walled pre-stressed wire-wound cylinder; the wire-winding and an internally water-cooled liner produce an effective heat exchanger between gas-gas and gas-water. Without this thin-wall solution, the cooling rates would be significantly lower and true rapid cooling could not be achieved. An extra advantage with rapid and uniform cooling is that the wear life of the furnace is increased dramatically, thereby dramatically lowering the costs associated with maintenance and spare/wear parts.

The gas cooling rates can be over 3,000°C/min, which is called uniform rapid quenching. The pressure is controlled and maintained through the full cooling time. After cooling, the temperature can increase again to heat treat the material and enable optimal grain growth before the final part cooling is done. Pressure is maintained during the whole cycle, which aids in achieving optimal grain size [5].

With this procedure, heat treatment becomes very advantageous in terms of softening, annealing and even tempering. The outcome is a better-quality material quality that results in reduced costs by reducing the scrap rate and shorter lead times due to less re-work. In addition, parts can be heat treated in the same furnace, thereby eliminating separate handling and additional steps for heat treatment (e.g., heating and subsequent quenching in water, oil or salt baths). This lowers the total capital investment, as well as the running costs.

## Material Properties

In many metal alloy systems, avoiding detrimental phases, such as the sigma phase in stainless steel or phase transformation between α- and β-phase in titanium, is crucial. By rapidly cooling the parts down into the safe region of the phase diagram, no detectable levels of these phases can be measured. Without rapid cooling, an increased level of mixed phases will negatively affect the material properties, grain growth and formation of oxides, carbides, and nitrides at the grain boundary [6].

Since ceramics are more fragile and far less ductile than metals, see Fig. 1, the uniform rapid cooling function is all about temperature control. For a ceramic part (e.g., nitrides, carbides, borides, spinels, etc.), having control of the cooling rate and the cooling curve is equally or more important than speed. The brittleness of the ceramic part makes it essential to make time for the cracking and re-stacking of ceramic powders in a Cold Isostatic Press (CIP) to move below 40% porosity. Sufficient time in the HIP is also necessary to form necking and pore closure, so the density can increase to close of the theoretical density (above 99% of theoretical density).

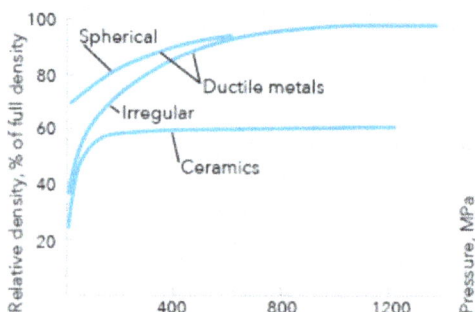

*Fig. 1. Density changes in powders.*

**Zirconia for Dental Implants [7]**
Zirconium ($ZrSiO_4$) belongs to the mineral group of silicates and was discovered in 1789 by the German chemist M. H. Klaproth. Zirconium dioxide (zirconia $ZrO_2$) is a natural compound of the element zirconium which occurs in nature. It has been used in dentistry for up to 25 years.

Zirconia has already been used over 40 years for industrial purposes. It is exceptionally durable and 100% biocompatible. For these reasons, it is used increasingly in surgery for ear, finger and hip prostheses. Applications for dentistry are found in zirconia pins, crowns, bridges and implants. The material's natural white base allows individual coloring in prescribed dentin shades. The biotechnical characteristics of zirconia result in high quality crowns, bridges and implants with excellent biocompatibility and aesthetic appearance.

The raw base material to produce zirconium dioxide is the mineral zirconium (ZrSiO4). Zirconium dioxide is produced in a chemical process using additives. Distinctions are made between sinter additives which influence the actual sinter process and characteristics of the final material, and auxiliary additives which facilitate workability. So-called "green blocks" are produced through a certified manufacturing process.

Sinter additives remain in the zirconium oxide, while the auxiliary additives (mainly volatile organic compounds apart from water) are removed from the material before sintering, leaving no residue. The green blocks are stabilized in a pre-sinter process to a degree that allows machining with tungsten milling tools.

Hot Isostatic Pressing – HIP'17                                        Materials Research Forum LLC
Materials Research Proceedings **10** (2019) 24-29          doi: http://dx.doi.org/10.21741/9781644900031-4

Framework milled from the chalky green blocks is enlarged by about 25%. Shrinkage during the final sinter fire at 1500°C results in the frame's final 1:1 size. The sinter process effects a shrinkage of 20%. Through it the frame acquires the final flexural strength and hardness by compaction of the material particles.

This is achieved by temperature-dependent diffusion processes with varying degrees of surface-, grain boundary- and volume diffusion. If the solidifying diffusion process happens too slow, sintering can be carried out under pressure in addition by hot isostatic pressing (HIPing) of the zirconia, see Fig. 2. Chemical composition and processing methods very much determine the quality characteristics of the end product.

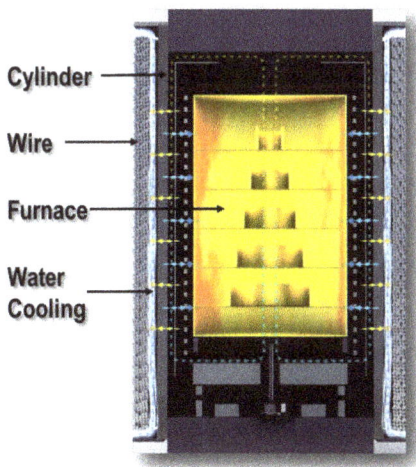

*Fig. 2. Illustration of a wire-wound HIP system with parts under operation.*

There is a difference between fully stabilized zirconia (FSZ) and partially stabilized zirconia (PSZ). Partial stabilization can be created by adding 3-6% CaO, MgO or $Y_2O_3$. Depending on production methods, the cubic, tetragonal or monocline modification can be stabilized. Partially stabilized zirconia displays high resistance to temperature changes which makes it suitable for use in an environment subject to high temperatures.

By adding 10-15% CaO and MgO, the cubic makeup of zirconia can be stabilized from absolute zero to fully stabilized zirconia. This makes the material thermally and mechanically resistant to temperatures up to 2600°C.

Due to low heat conductivity and a high thermal expansion coefficient fully stabilized zirconia has a lower resistance to temperature changes than partially stabilized zirconia. A composition of partially stabilized 95% $ZrO_2$ + 5% $Y_2O_3$ (5YSZ) presents the ideal material for dental restorations.

*Table 1. Typical composition of partially stabilized 5YSZ.*

| Chemical composition (%) | $ZrO_2$ $(+HfO_2)$ | $Y_2O_3$ | $Al_2O_3$ | $SiO_2$ | $Fe_2O_3$ | $Na_2O$ | Density (g/cm3) | Hardness (HV) |
|---|---|---|---|---|---|---|---|---|
| 5YSZ | Balance | 4.9-5.2 | 0.15-0.35 | < 0.02 | < 0.01 | < 0.04 | 6.02 | > 1200 |

**Productivity and Cost**

Cosiderably shorter cycle times in a HIP can be achieved with uniform rapid cooling. The obvious target is to increase productivity, which lowers the parts' cost by decreasing processing costs and investment depreciation time. The cycle time can be reduced by as much as 70% (see Fig. 3). For example, for a small- to medium-sized HIP unit, it is possible to run two cycles per 8-hr shift, instead of one cycle per shift (with natural cooling), and have time to achieve heat treatment in the HIP.

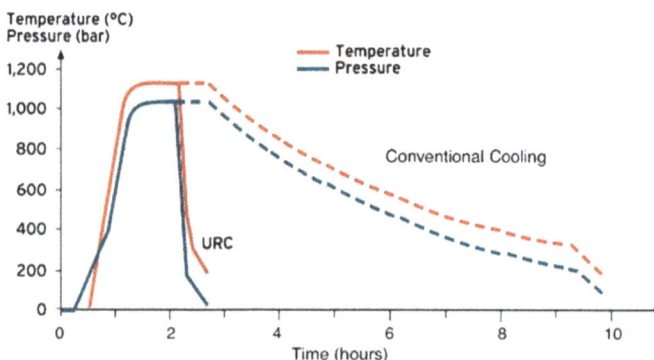

*Fig. 3. Typical HIP cycle times without and with uniform rapid cooling.*

The operating cost of HIPing has reduced drastically during recent years. This is due not only to uniform rapid cooling, but also to less maintenance-sensitive solutions, particularly of the furnace. For a large production HIP unit used for powder, or consolidation of large forgings and casting, the cost per kg HIPed material is around $0.20. This should be compared with the cost and time spent on NDT, X-rays, weld repairs and higher scrap rates, which to a large extent can be excluded when HIPing is used.

For partially stabilized 5YSZ, the main improvements after HIP is an increased density and drastically improved flexural strength. Typical values can be seen in table 2.

*Table 2. Material properties of partially stabilized 5YSZ, after sintering and after HIP.*

|           | Density (g/cm3) | Flexural Strength (MPa) |
|-----------|-----------------|-------------------------|
| Sintered  | 6.02            | 1250                    |
| HIPed     | 6.07            | 1725                    |

**Summary and Conclusion**

HIP technology provides manufacturers with opportunities to control their material properties and increase productivity. Productivity is increased two-fold through the use of uniform rapid cooling. In addition, combining HIP and heat treatment in the same equipment results in shorter lead times, better material properties, and processing and investment cost savings.

**References**

[1] Engineered Casting Solutions, pp. 30-34, Casting Source Directory, 2006.

[2] M. Diem, S. J. Mashl and R.D. Sisson, Heat Treating Progress, pp. 52-55, June/July, 2003

[3] A. Eklund, Powder Injection Moulding International, pp. 61-63, Vol. 6, No. 1, March 2012

[4] A. Eklund, M. Ahlfors, Metal Powder Report, Vol. 73, No. 3, pp. 163-169, May-June 2018. https://doi.org/10.1016/j.mprp.2018.01.001

[5] Angré, O. Karlsson, E. Claesson, "Phase Transformation under Isostatic Pressure in HIP", Proceedings from WorldPM2016, Hamburg, Germany, October 2016

[6] Heikkilä, J. Hagström, O. Karlsson, F. Gustavsson, "Phase Transformation in HIP – Dievar", Report No. KIMAB-2018-152.

[7] Klaproth, Martin Heinrich, Nordisk familjebok, 1st edition (1884).

# Effects of Hafnium on Microstructure and Mechanical Properties in as-HIPed FGH4097 Powder Metallurgy Superalloy

ZHANG Yi-Wen[1,2,a,*], JIA Jian[1,2], HAN Shou-Bo[1,2], LIU Jian-Tao[1,2]

[1]High Temperature Material Institute, Central Iron and Steel Research Institute, Beijing 100081, China

[2]Beijing Key Laboratory of Advanced High Temperature Materials, Beijing 100081, China

[a]yiwen64@126.com

**Keywords:** PM Superalloy, FGH4097, Hafnium, PREP, Mechanical Properties

**Abstract.** New type powder metallurgy (PM) superalloy FGH4097 developed in China is applied in advanced aircraft engine hot section components such as turbine disc and compressor disc. FGH4097 alloy billets ($\Phi 80 \times 135mm$) in the study were processed via plasma rotating electrode processing (PREP) powder making, hot isostatic pressing (HIP) forming and heat treatment. Range of powder size is 50~150μm. HIP parameters are 1180~1220℃, 130MPa and 4hrs. The heat treatment process of FGH4097 alloy billet is solution treatment plus three stages aging treatment. By means of metallurgical microscope, scanning electron microscope and physiochemical phase analysis, the influence of Hafnium with different contents on the grain size, $\gamma'$ phase, MC carbide in as-HIPed FGH4097 alloy have been studied, also influences of Hafnium contents on mechanical properties including tensile properties, stress rupture properties and fatigue crack propagation rate have been investigated. The results showed that Hf had no effect on grain size, $\gamma'$ phase size and MC carbide size and morphology. Hf promoted $\gamma'$ phase and MC carbide precipitation, changed $\gamma'$ phase and MC carbide chemical constitutions and accelerated the splitting of $\gamma'$ phase to preferably stable cubic. Proper Hf content plays a beneficial role in improving the comprehensive mechanical properties of FGH4097 alloy, which helps improve mechanical properties, including impact ductility, tensile plasticity, stress rupture life, and fatigue crack propagation resistance, also helps decrease notch sensitivity. FGH4097 alloy with 0.30wt%Hf content presents the optimum comprehensive mechanical properties.

## 1. Introduction

Microelement Hf added in Ni-based powder metallurgy (PM) superalloy can modify microstructure and improve stress rupture life, creep resistance, crack propagation resistance and eliminate notch sensitivity[1-4]. Microelement Hf promotes the precipitation of $\gamma'$ phase and MC carbide, which result in Hf contained $\gamma'$ phases and MC carbide precipitation, and their constitution changing[5,6]. Usually Hf is added to the composition of the advanced PM superalloys to change the mechanism of the MC carbide precipitation from the boundaries of grains and particles to inside the grains.

Previous research shown that Hf changed the morphology of $\gamma'$ phase in Ni-based PM superalloy. For example , Hf improved $\gamma'$ phase growth in IN100 PM alloy, cubic $\gamma'$ phase size in alloy contained 1.05%Hf (mass fraction; without notification, the following Hf content unit is mass fraction) was bigger than that contained 0.40%Hf, while Hf showed no apparent strengthening effect in alloys[1]. The addition of Hf in NASA IIB-11 PM alloy could improve latticed $\gamma'$ phase precipitation[5]. It is reported that adding 0.25-1.7%Hf in Astroloy PM alloy influenced morphology of $\gamma'$ phase apparently[6].With Hf content increasing, the $\gamma'$ phase arranges fan shape, which could be explained as one or more $\gamma'$ phases nucleated at grain boundary and grew in matrix. At the same time some elements diffused from matrix into $\gamma'$ phases to promote formation of $\gamma'$ phase branching, then $\gamma'$ phase grows radicaly and arranges fan shape[6]. With appropriate Hf content addition in Ni-based PM superalloy, MC carbide precipitation distributed dispersally[1,2,6]. Under the action of elastic strain field, the cubic $\gamma'$ phase splitted into other morphology, like doublet of plantes or octet of cubes. And the size of $\gamma'$ phase will be smaller[7-10]. In summary, the morphology of $\gamma'$ phase will be different in differrent PM superalloy with Hf content.

IN100 alloy with 0.4%Hf content (MERL76 alloy) prepared by argon atomization and hot isostatic pressing (HIP) showed that the stress rupture life of MERL76 alloy was ten times longer than that of IN100 alloy at the condition of 732℃ and 655 MPa. The stress rupture ductility was much better and there was no notch sensitivity in MERL76 alloy[1]. EP741 alloy with 0.3%Hf content (EP741NP alloy) prepared by plasma rotating electrode process (PREP) atomization and HIP showed that both stress rupture life and ductility at 750°C were improved[2-4]. Because Hf forms primary MC carbides that remain stable at low temperatures, which is considered to prevent excessive $M_{23}C_6$ and $M_6C$ formation at grain boundaries, RR1000 alloy[11,12](0.75%Hf), N18 alloy[12,13](0.5%Hf) and N19 alloy[14](0.25%Hf) showed improved creep resistance and crack propagation resistance.

The addition of too much Hf has little effect on improving mechanical properties of Ni-based PM superalloy[2]. Till now the content of Hf is lower than 0.8% in N18, RR1000 and EP741NP alloys[4,12,15]. Investigations about the effect of Hf addition on mechanical properties in PM superalloys have not been addressed systematically in literatures. Thus, this paper investigated the effects of Hf contents (0, 0.16, 0.30, 0.58 and 0.89%) on chemical constitution, morphology, size and volume fraction of $\gamma'$ phase and MC carbide. The effects of Hf contents on impact ductility, tensile properties(at room temperature, 650℃ and 750℃), stress rupture properties (at 650℃) and fatigue crack growth rate (at 650℃) have also been studied. According to the results, the reasonable Hf content was given. This work is helpful to understand Hf effects on mechanical properties in PM superalloy and to design Ni-based PM superalloy with high strength as well as high damage tolerance.

1 Materials and methods

The experimental material was Ni-based PM superalloy FGH4097 with different Hf contents. The chemical composition (mass fraction, %) of FGH4097 is: Co16.0, (Cr+W+Mo) =18.5, (Al+Ti+Nb)=9.4, Hf 0-0.89, C minor, B minor, Zr minor, and Ni balance. The contents of Hf (mass fraction, %) were 0、0.16、0.30、0.58 and 0.89. Powders (50-150 μm) prepared by PREP and HIPed at 1200 °C. HIP billets heat treatment process was 1200°C/4 h/AC + three stages aging treatment. The last aging treatment was 700 °C/(15-20)h/AC.

The chemical constitution and contents of $\gamma'$ phase and MC carbide phase have been analyzed by physiochemical phase analysis. Carbide morphology has been observed by JSM-6480L scanning electron microscope(SEM), size of carbide has been analyzed by image analyzer (ten times at 500x). Microstructure of $\gamma'$ phase has been observed by JSM-6480L SEM and SUPRA 55 FEG-SEM. SEM samples for microstructure observation were prepared by electrolytic polishing and electrolytic erosion. The electrolytic polishing was carried out in 20%$H_2SO_4$+80%$CH_3OH$ solution, at 30V and 15-20s, the electrolytic polishing erosion was carried out in 85ml$H_3PO_4$+5ml$H_2SO_4$+8g$CrO_3$ solution, at 5V and 3~6s respectively. Size of $\gamma'$ phase has been analyzed by Image-Pro Plus 6.0 software. The impact ductility, tensile

Materials Research Forum LLC
doi: http://dx.doi.org/10.21741/9781644900031-5

properties(at room temperature, 650℃ and 750℃) have been tested. Stress rupture life at 650℃ /1020MPa has been tested at smooth and notched samples(radius of notch R=0.15mm). The fatigue crack growth rate at 650℃ has been tested under condition of stress ratio R=0.05 and load frequency 10-30 cycle/min. The samples were prepared by mechanical polishing and chemical etching (etched by $3gCuSO_4+80mlHCl+40ml$ $C_2H_5OH$ solution). After chemical etching, $\gamma'$ phase is etched away and $\gamma$-matrix is left behind, then these samples have been used to micro-hardness test in MTS XP Nano Indenter(three points in one sample). The effect of Hf content on micro-hardness of $\gamma$-matrix in FGH4097 alloy has been investigated.

## 2. Experimental results

### 2.1 Phases in FGH4097 alloy

The experimental results showed that FGH4097 alloy consists of $\gamma$ matrix, $\gamma'$ phase, MC carbide, small amounts of $M_6C$ carbide and $M_3B_2$ boride, and $\gamma'$ phase and MC carbide are the main precipitates. The results of physicochemical phase analysis of the phase content in FGH4097 alloy with different Hf contents was given in Table 1. With the increase of Hf content, the amount of MC carbide increases and the amount of $\gamma'$ phase also increases slightly. The amount of $\gamma'$ phase is about 62%, the amount of MC carbide is no more than 0.34%, and the total amount of $M_6C$ carbide and $M_3B_2$ is no more than 0.21%. Thus, no new phase was found in FGH4097 alloy with different Hf contents. The grain shapes were regular, the grain sizes were 30 ~ 40μm which keeps in a stable level in FGH4097 alloy when Hf contents changed from 0 to 0.89%.

Table 1 The amount of phases in FGH4097 alloy with different Hf contents (mass fraction, %)

| Hf content | $\gamma$ | $\gamma'$ | MC | $M_6C+M_3B_2$ |
|---|---|---|---|---|
| 0 | 37.678 | 61.930 | 0.264 | 0.128 |
| 0.16 | 37.503 | 62.080 | 0.266 | 0.151 |
| 0.30 | 37.378 | 62.180 | 0.270 | 0.172 |
| 0.58 | 37.062 | 62.450 | 0.293 | 0.195 |
| 0.89 | 36.762 | 62.690 | 0.338 | 0.210 |

### 2.2 Morphology and chemical constitution of $\gamma'$ phase

In FGH4097 alloy with five Hf contents, $\gamma'$ phase distributed in grains and at grain boundaries. There are three kinds of $\gamma'$ phase: the large $\gamma'$ phase precipitated from $\gamma$ solid solution during solution cooling which is called primary $\gamma'$ phase at grain boundaries, square $\gamma'$ phase precipitated in grains precipitated from $\gamma$ solid solution during solution cooling which is called secondary $\gamma'$ phase, fine and square $\gamma'$ phase precipitated in grains from $\gamma$ solid solution during aging which is called ternary $\gamma'$ phase. Figure 1 showed the morphology of $\gamma'$ phase in FGH4097 alloy with 0.30% Hf content, $\gamma'$ phase at grain boundaries and secondary $\gamma'$ phase are block shapes (Fig.1a) and ternary $\gamma'$ showed spherical particles (Fig.1b).

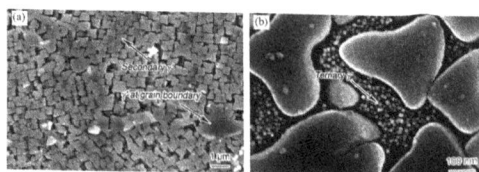

Fig.1 Low (a) and high (b) magnified SEM images of $\gamma'$ phase in FGH4097 alloy with 0.30% Hf

The experimental results showed that the addition of small amount of Hf in FGH4097 alloy does not change morphology of $\gamma'$ phase at grain boundaries and ternary $\gamma'$ phases, but greatly influence morphology of secondary $\gamma'$ phase. Figure 2 showed SEM images of secondary $\gamma'$ phase in FGH4097 alloy with different Hf contents. The secondary $\gamma'$ phase is mainly cubic in the alloys without Hf and with 0.16wt% Hf contents(Fig. 2a, b). As the amount of Hf increases, the secondary $\gamma'$ phase grows and splits. For FGH4097 alloy with 0.30% Hf content, the secondary $\gamma'$ phase is mainly octahedron cubic and butterfly-like shape (Fig. 2c). For FGH4097 alloy with 0.58% Hf content, the secondary $\gamma'$ phase is mainly cubic and octahedron cubic (Fig. 2d). The secondary $\gamma'$ phase is mainly cubic in the alloy with 0.89% Hf content (Fig. 2e).

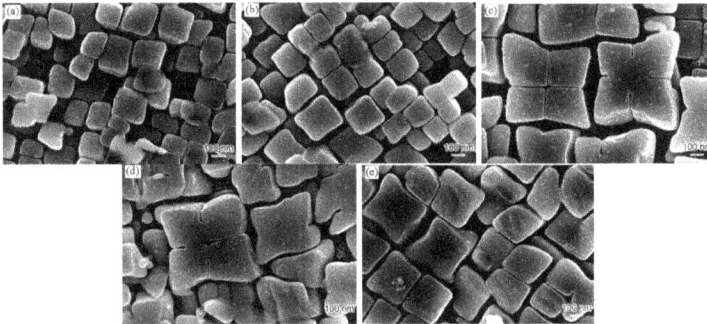

*Fig.2  SEM images of secondary $\gamma'$ phase in FGH4097 alloys with 0 (a), 0.16% (b), 0.30% (c), 0.58% (d) and 0.89% (e) Hf respectively*

When Hf element was added into FGH4097 alloy, because of Hf replacing Al, the chemical constitution of $\gamma'$ phase is $(Ni,Co)_3(Al,Ti,Nb,Hf)$ which contain Ni, Co, Al, Ti, Nb and Hf. With the increase of Hf content in the alloy, Hf content in $\gamma'$ phase increases gradually, while Al content decreases gradually, and more amount Al replaced by Hf. When Hf content is 0, 0.16, 0.30, 0.58 and 0.89%, chemical constitution of $\gamma'$ phase are $(Ni_{0.852}Co_{0.148})_3(Al_{0.783}Ti_{0.129}Nb_{0.088})$, $(Ni_{0.854}Co_{0.146})_3(Al_{0.781}Ti_{0.129}Nb_{0.088}Hf_{0.002})$, $(Ni_{0.855} \quad Co_{0.145})_3(Al_{0.778}Ti_{0.129}Nb_{0.088}Hf_{0.005})$, $(Ni_{0.856}Co_{0.144})_3(Al_{0.773}Ti_{0.129}Nb_{0.088}Hf_{0.010})$ and $(Ni_{0.857} \quad Co_{0.143})_3(Al_{0.767}Ti_{0.129}Nb_{0.088}Hf_{0.016})$, respectively. With the increase of Hf content in the alloy, size of ternary $\gamma'$ phase keeps in a stable level and average size is 14-18nm. Sizes of $\gamma'$ phase at grain boundaries and secondary $\gamma'$ phase increase firstly and then decrease gradually. Average size of $\gamma'$ phase at grain boundary is 820-1450nm and that of secondary $\gamma'$ phase is 276-511nm. Sizes of $\gamma'$ phase at grain boundaries and secondary $\gamma'$ phase in 0.30% Hf content FGH4097 alloy have the maximum size values which are average 1450nm and 551nm, respectively. Size of secondary $\gamma'$ phase in FGH4097 alloy with different Hf contents showed a normal distribution.

## 2.3  Morphology and chemical constitution of MC carbide

Addition of Hf in FGH4097 alloy did not change the morphology of MC carbide. Granular MC carbide distributed in grains and at grain boundaries, MC carbide precipitated at grain boundaries has larger size. With the increase of Hf content in FGH4097 alloy, Size of MC carbide keeps in a stable level and the average size is $0.878\sim1.064\mu m$. Figure 3 showed SEM images of MC carbide in FGH4097 alloy with 0.30% Hf content.

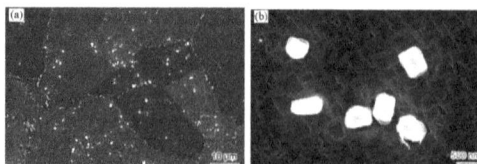

*Fig.3 Low (a) and high (b) magnified SEM images of MC carbide of FGH4097 alloy with 0.30% Hf*

When Hf is added in FGH4097 alloy, MC carbide chemical constitution is (Nb, Ti, Hf)C, which mainly contained Nb, Ti, Hf and C. When the content of Hf is 0, 0.16, 0.30, 0.58% and 0.89%, chemical constitutions of MC carbide are $(Nb_{0.664}Ti_{0.336})C$, $(Nb_{0.654}Ti_{0.323}Hf_{0.023})C$, $(Nb_{0.642}Ti_{0.308}Hf_{0.050})C$, $(Nb_{0.619}Ti_{0.280}Hf_{0.101})C$ and $(Nb_{0.574}Ti_{0.253}Hf_{0.173})C$, respectively.

## 2.4 Mechanical properties

### 2.4.1 Impact toughness

The experimental results showed that addition of Hf apparently influences impact toughness at room temperature. Figure 4 showed the impact energy absorption at room temperature in FGH4097 alloy with different Hf contents. With the increase of Hf content, impact absorption increased at first and then decreased. Impact energy absorption value of the alloy with 0.16% Hf content is the highest, that of the alloy with 0.30wt% Hf content is the second highest one and that of the alloy with 0.89% Hf content is the lowest one. Compared with 0.30% Hf content, impact energy absorption values of the alloys without Hf and with 0.89% Hf content alloy were reduced by 11.2% and 21.0% respectively, the impact toughness decreased obviously.

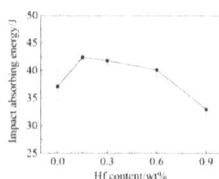

*Fig.4 Impact absorbing energy of FGH4097 alloy with different Hf contents at room temperature*

### 2.4.2 Tensile properties

Tensile properties of FGH4097 alloy with different Hf contents at room temperature has been shown in Fig.5. Hf content has little effect on the tensile strength and yield strength at room temperature. Room temperature tensile strength of 0.16% Hf content FGH4097 alloy is slightly higher, and 0.30% Hf content FGH4097 alloy has the lowest tensile strength at room temperature. Tensile ductilitychanges with the increase of Hf content. Tensile ductilityof 0.16% Hf content FGH4097 alloy is the lowest, while tensile ductilityis the highest when Hf content is 0.30%.

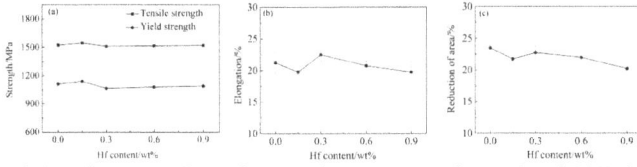

*Fig.5  Strength (a), elongation (b), reduction of cross sectional area (c) of FGH4097 alloy with different Hf content at room temperature*

Tensile test of FGH4097 alloy with different Hf contents at 650℃ has been shown in Fig.6. with Hf content increasing, curves of tensile strength changing at 650℃ is similar to those at the room temperature. Hf content has little effect on tensile strength and yield strength, but influence tensile ductilityto some degree. When Hf content is 0.30%, tensile strength is the lowest and ductilityis the highest. When Hf content is 0.16%, tensile strength is the highest and ductilityis the lowest.

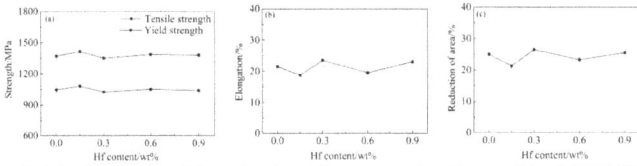

*Fig.6  Strength (a), elongation (b), reduction of cross sectional area (c) of FGH4097 alloy with different Hf content at 650 ℃*

Influence of Hf content on tensile properties of FGH4097 alloy at 750℃is the same as that at 650℃. Tensile properties with different Hf contents in FGH4097 alloy at 750℃ is shown in Fig.7. Hf content has little effect on tensile strength and yield strength, while has more effect on the tensile plasticity. When Hf content is 0.30%, tensile strength is the lowest and ductilityis the highest. When Hf content is 0.16%, tensile strength is the highest and ductilityis the lowest.

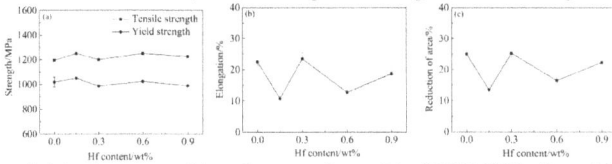

*Fig.7  Strength (a), elongation (b), reduction of area (c) of FGH4097 alloy with different Hf contents at 750 ℃*

### 2.4.3  Stress rupture properties at high temperature

Under 650℃/1020MPa test condition, stress rupture life of smooth sample and notch sample (R=0.15mm) in FGH4097 alloy with different Hf contents were shown in Figure 8. Stress rupture life of notch specimens increased with Hf content increasing. Stress rupture life of notch specimen with 0.30% Hf content is the longest one. Stress rupture life of notch specimen with no Hf content is shorter than that of smooth specimen and notch sensitivity exists. When Hf content is higher than 0.16%, stress rupture life of notch specimen is longer than that of smooth specimen and notch sensitivity does not exist. The law of rupture ductility with Hf content

Hot Isostatic Pressing – HIP'17

Materials Research Forum LLC

Materials Research Proceedings **10** (2019) 30-38

doi: http://dx.doi.org/10.21741/9781644900031-5

changing is the same as the tensile ductilityin FGH4097 alloy, rupture ductilityof the alloy with 0.30% Hf content is the highest value.

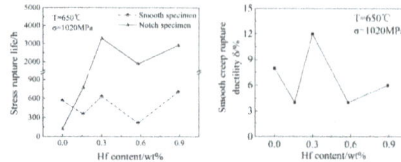

*Fig.8 Stress rupture life of FGH4097 alloy with different Hf contents at 650 ℃/1020MPa*

### 2.4.4 Fatigue crack propagation rate

Figure 9 showed the relationship between fatigue crack propagation rate (da/dN) and stress intensity factor range ($\Delta K$) in FGH4097 alloy with different Hf contents under condition of 650 ℃/R=0.05/0.33Hz. The da/dN of FGH4097 alloy with 0.30% Hf content is the lowest during the $\Delta K$ range of 28.5~33.2 MPa•m$^{1/2}$. When $\Delta K$=30MPa•m$^{1/2}$, the fatigue crack propagation rate values of the alloys with 0, 0.16, 0.30, 0.58 and 0.89%Hf content are $2.50\times10^{-3}$, $1.47\times10^{-3}$, $4.85\times10^{-4}$, $1.04\times10^{-3}$ and $1.04\times10^{-3}$mm/cycle, respectively. The da/dN value of the alloy with 0.30% Hf content is one-fifth of that of the alloy without Hf content.

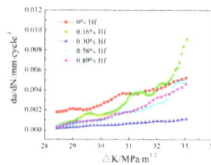

*Fig.9 Fatigue crack propagation rate of FGH4097 alloy with different Hf contents at 650 ℃*

### 3. Discussions

It can obviously improve the ductilitywhen small amount of Hf is added in FGH4097 alloy. Especially, mechanical properties are obviously improved when Hf content is 0.30%. Hf addition changes distribution of alloy elements between phases and affects microstructure and properties. Notch sensitivity elimination is strongly dependent on ductility of γ matrix solid solution. Hf is a strong carbide-forming element. Hf partitions to MC carbide firstly and then to γ′ phase. Hf modifies the chemical constitutions of MC phase, γ′ phase and γ matrix, which led to redistribution of elements Ti, Nb, W, Mo, and Cr in MC, γ′ phase and γ matrix.

When Hf is 0.3% content, Hf can effectively replace Nb and Ti in MC. The displaced Nb and Ti partition to γ matrix and form NbC and TiC respectively in γ matrix. As a result, C content in γ matrix decreases which decreases γ matrix strength and improves ductility of the alloy. Particularly, although the tensile strength decreased slightly, the strength matched well with ductility. It can also decrease crack propagation rate, improve impact toughness, eliminate notch sensitivity of stress rupture. If Hf content is not enough, Hf can not effectively replace Nb and Ti in MC, the strength of γ matrix will not decrease efficiently. If Hf content is too much, Hf will form HfC or $HfO_2$ and hardly replaces Nb and Ti in MC. The results by MTS Nano Indenter XP showed that the hardness of γ matrix in alloy with 0.30%Hf content is the lowest which means γ matrix strength is the lowest. Thus, the amount of Hf in FGH4097 alloy should be added properly.

## 4. Conclusions

(1) Precipitation phases of FGH4097 alloy with different Hf content are mainly $\gamma'$ phase, MC carbide and small amount of $M_6C$ carbide as well as $M_3B_2$ boride. When Hf is added in FGH4097 alloy, part of Hf replace Al atoms in $\gamma'$ phase, constitution of $\gamma'$ phase is $(Ni,Co)_3(Al,Ti,Nb,Hf)$; part of Hf replace Ti and Nb atoms in MC type carbide, constitution of MC carbide is $(Nb,Ti,Hf)C$. As Hf contents increasing, $\gamma'$ phase fraction slightly increased and MC carbide fraction increases apparently. Hf addition in FGH4097 alloy does not influence size and morphology of MC carbide, but significantly affects size and morphology of $\gamma'$ phase. Hf enters into $\gamma'$ phase and changes the elastic interaction energy distribution of $\gamma'$ phase during growth, which accelerate cubic $\gamma'$ phase splitting into 8 small cubic and promote $\gamma'$ phase forming stable cubic state preferably. Since the addition of Hf effects the distribution of alloy elements between phases, thereby improving the ductility of the $\gamma$ matrix solid solution with the optimal Hf, which is beneficial for reducing the notch sensitivity.

(2) Hf content does not influence tensile strength of FGH4097 alloy significantly, but influence its impact toughness and tensile ductility obviously. Under $650°C/1020MPa$ condition, stress rupture life of notch sample is shorter than that of smooth one when there is no Hf addition. Hf addition decreases the notch sensitivity of sample, proper Hf addition can improve the fatigue crack propagation resistance.

(3) The best comprehensively mechanical properties including tensile strength, stress ruputure property, plasticity, toughness and crack propagation resistance can be reached in 0.30%Hf content FGH4097 alloy.

(4) Proper Hf addition in FGH4097 alloy can effectively change redistribution of elements between precipitation and $\gamma$ solid solution which means that the strength and ductilitycan have a good match. It can also decrease notch sensitivity of high temperature stress rupture and crack propagation rate.

(5) To significantly explain the mechanism of effect of Hf on the mechanical properties in FGH4097 alloy, more in-depth research and analysis are needed.

## Acknowledgements

The authors would like to thank the financial support by International Science & Technology Cooperation Program of China (No 2014DFR50330).

## References

[1] Evans D J, Eng R D. Development of a high strength hot isostatically pressed(HIP) disk alloy, MERL76, Hausner H H, Antes H W, Smith G D. Modern Developments in Powder Metallurgy. Washington: MPIF and APMI. 1982, Vol. 14, 51-63.

[2] Belov A F, Anoshkin N F, Fatkullin O Kh, et al. Features of alloying of high-temperature alloys obtained by the metallurgical method of granules, Bannikh O A. Heat-resistant and Zharostoye Steels and Alloys on the Nickel Base. Moscow: Science, 1984, pp. 31-40.

[3] Radavich J, Carneiro T, Furrer D, et al. Effect of processing and composition on the structure and properties of P/M EP741NP type alloys, Chinese Journal of Aeronautics, 20 (2007) 97-106. https://doi.org/10.1016/S1000-9361(07)60013-2

[4] Radavich J, Furrer D, Carneiro T, et al. The microstructure and mechanical properties of EP741NP powder metallurgy disc material, Reed R C, Green K A, Caron P, et al(Eds.), Superalloys 2008, TMS, Pennsylvania, 2008, pp. 63-72.

[5] Miner R V. Effects of C and Hf concentration on phase relations and microstructure of a wrought powder-metallurgy superalloy, Metallurgical Transactions A, 8 (1977) 259-263. https://doi.org/10.1007/BF02661638

[6] Larson J M, Volin T E, Larson F G. In: Braun J D, Arrowsmith H W, McCall J L. Microstructural Science, American Elsevier Pub, New York, 1977, pp. 209-217.

[7] Wang Y, Chen L Q, Khachaturyan A G. Kinetics of strain-induced morphological transformation in cubic alloys with a miscibility gap, Acta Metallurgica et Materialia, 41 (1993) 279-296. https://doi.org/10.1016/0956-7151(93)90359-Z

[8] Qiu Y Y. Retarded coarsening phenomenon of $\gamma'$ particles in Ni-based alloy. Acta Materialia, 44 (1996) 4969-4980. https://doi.org/10.1016/S1359-6454(96)00128-0

[9] Yoo Y S, Yoon D Y and Henry M F. The effect of elastic misfit strain on the morphological evolution of $\gamma'$- precipitates in a model Ni-base superalloy, Metals and Materials, 1 (1995) 47-61.

[10] Qiu Y Y. The splitting behavior of $\gamma'$ particles in Ni-based alloys, Journal of Alloys and Compounds, 270 (1998) 145-153. https://doi.org/10.1016/S0925-8388(98)00462-9

[11] Hardy M C, Zirbel B, Shen G. Developing damage tolerance and creep resistance in a high strength nickel alloy for disc applications, Green K A, Pollock T M, Haradra H, et al (Eds.), Superalloys 2004, TMS, Pennsylvania, 2004, pp. 83-90.

[12] Starink M J, Reed P A S. Thermal activation of fatigue crack growth: Analysing the mechanisms of fatigue crack propagation in superalloys, Materials Science and Engineering A, 491 (2008) 279-289. https://doi.org/10.1016/j.msea.2008.02.016

[13] Wlodek S T, Kelly M, Alden D. The structure of N18.Antolovich S D, Stusrud R W, Mackay R A, et al(Eds.), Superalloys 1992, TMS, Pennsylvania, 1992, pp. 467-476.

[14] Guédou J-Y, Augustins-Lecallier I, Nazé L, et al. Development of a new fatigue and creep resistant PM nickel-base superalloy for disk applications. Reed R C, Green K A, Caron P, et al (Eds.), Superalloys 2008, TMS, Pennsylvania, 2008, pp. 21-30.

[15] Flageolet B, Villechaise P, Jouiad M, Mendez J. Ageing characterization of the powder metallurgy superalloy N18. Green K A, Pollock T M, Haradra H (Eds.), Superalloys 2004, TMS, Pennsylvania, 2004, pp. 371-379.

Hot Isostatic Pressing – HIP'17                                    Materials Research Forum LLC
Materials Research Proceedings **10** (2019) 39-46          doi: http://dx.doi.org/10.21741/9781644900031-6

# Development of the Design and Technological Solutions for Manufacturing of Turbine Blisks by HIP Bonding of the PM Disks with the Shrouded Blades

Liubov Magerramova[1,a], Eugene Kratt[2,b]

[1] Soviet Army st., 3-94, Moscow 127018, Russia (Central Institute of Aviation Motors, Moscow 111116, RF)

[2] G. Kurina st., 42-114, Moscow, 121108, Russia (Laboratory of New Technology, Moscow, RF)

[a]lamagerramova@mail.ru, [b]kratt@rambler.ru

**Keywords:** Gas Turbine Engine, Hot Isostatic Pressing, Bimetallic Blisk, Nickel Superalloy

**Abstract.** A bimetallic turbine blisk with shrouded blades had been designed in order to enhance the gas-dynamic and strength characteristics of the wheels of the gas turbine engines and reduce their weight. The separately cast nickel superalloy shrouded blades are joined to disc made of the heat-resistant alloy powder using a method of hot isostatic pressing (HIP). The problem of such joint is complicated by the presence of the shrouds on the periphery of the blades. This design should provide a good contact on the working faces of the shrouds during the operation. In order to solve this problem, a capsule for manufacturing of the disk piece and a process flow diagram for calculation of shaping such a capsule during hot isostatic pressing have been developed.

## Introduction

In various fields of technology, the need for high performance parts and components is constantly increasing. The shape and dimensions of blanks for such parts need to be close to the geometry of the final parts (the near net shape for the disk and net shape geometry for the blades). Traditional technology based on casting or forging and machining have serious limitations in the production of such blanks, due to the considerable difficulties in ensuring the requirements of geometrical complexity and required accuracy and distribution of performance and technological characteristics in the material.

The development of new technologies for the creation of bimetallic wheel structures, based on the processes of diffusion bonding of dissimilar alloys now allows us to meet the extremely high (and increasingly stringent) requirements for safe operation and economic efficiency of Gas Turbine Engines (GTE). Various methods are used to connect the blades to the disks; for example, electrochemical and mechanical methods of disk manufacture with blades from one forging, or blade connection to the disc by indentation at high temperature, soldering, friction welding, or hot isostatic pressing [1-4].

During the recent decades worldwide a  progressive technological method based on Hot Isostatic Consolidation of rapidly solidified powders of advanced alloys into various shapes has been developed.

Hot isostatic pressing (HIP) of powder materials is a process of high-temperature consolidation of porous workpieces under a high pressure. This process eliminates the above-mentioned restrictions, allowing obtaining a workpiece of the required complex configuration. Such parts as impellers, diffusers, turbines are produced by hot isostatic pressing. Hot isostatic pressing is advantageous due to its ability to produce complex shape parts earlier available only by casting from difficult to process alloys, and possessing mechanical properties of the wrought materials.

In the variety of technological parameters influencing the final shape of the product the decisive one is the design of capsule as a deformable tool that determines the original shape of the workpiece and transmitted pressure. From the point of view of HIP, presence of a deformable capsule leads to a nonuniform stress in the powder material and the complex change of the bulk shape [5, 6].

The complication of process within a wide range of pressures and temperatures and the absence of rigid forming tool require constant analysis (forecasting the ultimate shape of the workpiece after HIP) and synthesis (designing capsules ensuring the desired shape and size). Further improvement of HIP technology is associated with an increase of quality and reliability aimed at effective manufacturing of parts with surfaces that do not require additional machining.

### The requirements to the turbine wheel of a gas turbine engine

The wheel of a high-pressure turbine engine is a key component that defines the main engine characteristics. A reasonable approach to increasing the efficiency of turbine of such type is to use blade shrouding. However, this shrouding also increases the centrifugal loading on the profile part of the blades, the lock connection, and the disk.

The wheel elements of gas turbines operate under of non-uniform loading and radial heating. Hence, the materials for the different parts of the wheel should meet different requirements. Therefore, the wheel should combine the materials satisfying the required working conditions in the appropriate areas.

One area recognized as a stress concentrator is the lock connection of the blades with the disk. The stress at this location limits the life of the wheel, and its reinforcement results in an increased disk weight. Wheels with detachable blades often do not allow placing of the necessary number of blades from the point of view of gas-dynamic efficiency.

One solution of this problem is to eliminate the lock connections. The design of the wheel may take the form of a Bimetallic Blisk (BB), where the blades and the disk part are made of different materials and HIP bonded together [7 - 9]. Several blisks made of heterogeneous alloys for small turbines were designed in CIAM [10 - 14].

The shrouds are necessary for improved efficiency as they reduce the gas flow in the radial gap. In general, shrouds are designed for damping and detuning purposes, especially of long thin blades. These goals are achieved by designing the working wheels in such a way that they ensure both a reliable contact on the shroud working faces and an acceptable level of stress, especially in thin peripheral blade sections. The blades are assembled with interference between the working faces to ensure the contact between the blade shrouds [7].

The technological design of the turbine wheel should meet the following requirements:
- Reduction of the wheel mass,
- Increasing of the turbine wheel efficiency,
- Ensuring of reliable contact of the shroud working faces during operation,
- Meeting of strength and detuning requirements,
- The possibility of manufacturing.

The designed BB consists of heat resistant superalloy shrouded blades bonded to a nickel alloy powder disk. A single-stage wheel turbine for small-sized engine was designed with the goal of improving its efficiency. The blades of blisks were proposed to be made with shrouds in order to minimize the losses due to the radial clearance at the shroud (Fig. 1a).

Hot Isostatic Pressing – HIP'17                                    Materials Research Forum LLC
Materials Research Proceedings **10** (2019) 39-46            doi: http://dx.doi.org/10.21741/9781644900031-6

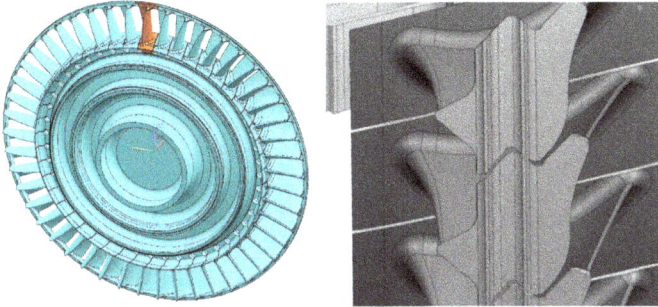

*Figure 1. a)The project of BB with shrouds, b)The shrouds*

The shroud construction is aimed at minimizing its weight and meets the above mentioned requirements. As calculation has shown, the efficiency of the turbine with shrouded blades is increased by 0.2% compared to that of a wheel without shroud.

The preliminary gap between the working edges of the shrouds (Fig. 1b) should be provided due to the peculiarities of the technological process of manufacturing blisks [8].

The location and configuration of the connection zone of the blade with the disc part are determined by the demand of minimization of weight of wheels and satisfaction of the requirements of strength under specified conditions of operation and capabilities of the production process.

The cross section of the shank of the blade is parallelogram shaped. Such shank is required to ensure the specified angles of the vanes in the gas channel (Fig. 2).

Initially the optimal position of the connection zone of the wheel is determined on the basis of equal strength of materials of the blades and disk part with respect to temperature in this area on the most damaged mode. Then the location of the connection zone is adjusted from the design considerations.

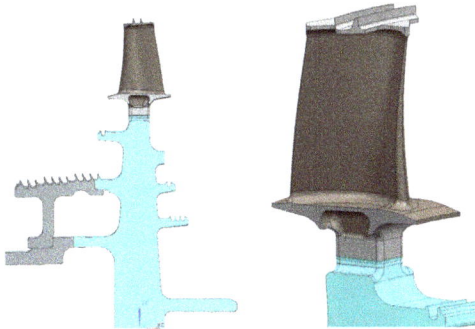

*Figure 2. The connection zone of the blades with the disc part*

The area of the connection zone is pre-determined by the satisfaction of requirements of safety for bearing capacity. On the basis of computation, the blisk turbine was designed with the weight reduced by ~10% compared to the original wheel design with a fir-tree connection.

**Technological aspects**

Development of technological process of manufacturing BB turbine with shrouded blades includes the following steps:

- Development of sonic shape blanks for BB with the features of NDT methods (fig.3),

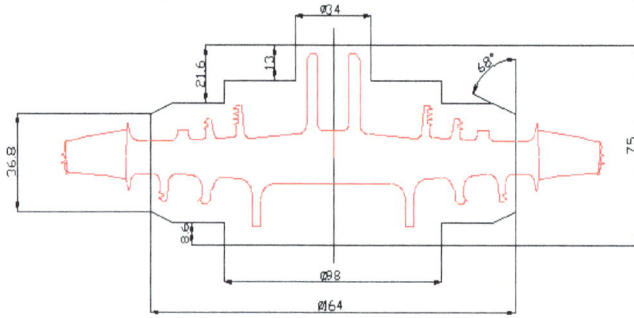

*Figure 3. Sonic shape blanks for BB*

- Designing of the HIP tool (Capsule) for the manufacture of the blank of the BB with shrouded blades,
- FEM Modeling of deformation of the capsule during the process of hot isostatic pressing,
- Development of the drawings for the capsule,
- Development of manufacturing technology for the capsule elements.

The problem of bonding the blades to the disc part to form a single part using the HIP method is as follows. The process of joining these components at high temperature and pressure is accompanied by the shrinkage of the powder disc alloy in the gap between the disc part and the blades. In this case the blades move in the direction of the disk. However, the blades cannot move if there is no clearance between the blade shrouds. In this case, good bonding of the shanks of the blades with the disc part is more difficult to achieve.

Allowing movement of the blades to the disk in the process of shrinkage of the powder requires the assembly of the blades in the capsule in a special fixture that provides gaps between the shrouds. The value of these gaps must be reduced to zero after successful completion of the connection of the blades with the disc.

For the manufacturing of Blisk by hot isostatic pressing, the workpiece forming simultaneous connection of hard disk and blade parts in the capsule filled with powder was itself a special forming tool (Fig. 4). The capsule was filled with powder material (in these case alloy EP 741NP). The cast alloy blades GS32 and shaping steel inserts are inside the capsule. Inserts that are consequently removed from the finished product leave the cavity of the desired shape. During HIP capsule undergoes uneven shrinkage, taking the form of the product. Precisely the complexity of modeling of shrinkage for this 3-dimensional object is the major difficulty in obtaining products of required geometry.

Hot Isostatic Pressing – HIP'17                                    Materials Research Forum LLC
Materials Research Proceedings **10** (2019) 39-46          doi: http://dx.doi.org/10.21741/9781644900031-6

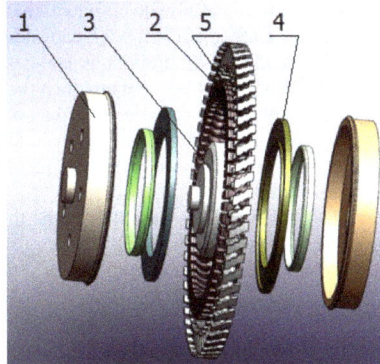

*Figure 4. Capsule for forming a BB: 1-capsule, 2- inserts, 3-central part, 4-fixing elements, 5-cast blades*

The task of the mathematical description of the process is to design the capsule and inserts in such a way that the final shape of the powdered products had the required geometry. The advantage of analytical descriptions is that it allows to more fully analyze the dependence of the required characteristics of the parameters of process. Since this task is physically and geometrically nonlinear, such precision is difficult to achieve: first, purely mathematical difficulties; secondly, the main difficulty is defining relations precisely describing the process and the rheological properties of the processed materials.

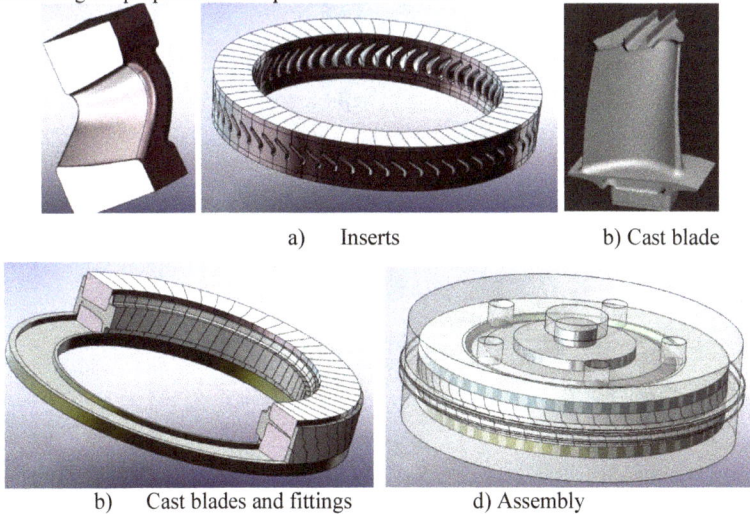

a)     Inserts                                        b) Cast blade

b)     Cast blades and fittings          d) Assembly

*Figure 5. View of the individual capsule elements and their assembly*

In practice, the process of manufacturing of powder products is an iterative process, and its essence is as follows: construction of a mathematical model, capsule design based on this model, manufacturing of the product. The geometry of product is compared with the required, then basing on this mapping, the mathematical model is tuned and then the process is repeated again.

Figure 5 shows the individual elements of the capsule as well as snap-in assembly sequence of the capsule.

The mathematical modeling of HIP process is of particular interest, as the ultimate shape of the final product depends on the geometry of the capsules.

Acceptable mathematical model of HIP process should meet the following requirements.

1. It gives a dimensionally close first approximation;

2. It properly takes into account the influence of process parameters and rheological properties of the materials;

3. It allows making changes in the model parameters basing on the results of the experiment, and, if necessary, entering additional parameters.

The work was produced by the FEM Modeling of deformation of the capsule with powder, inserts and blades during hot isostatic pressing. The simulation results are shown in Fig. 6.

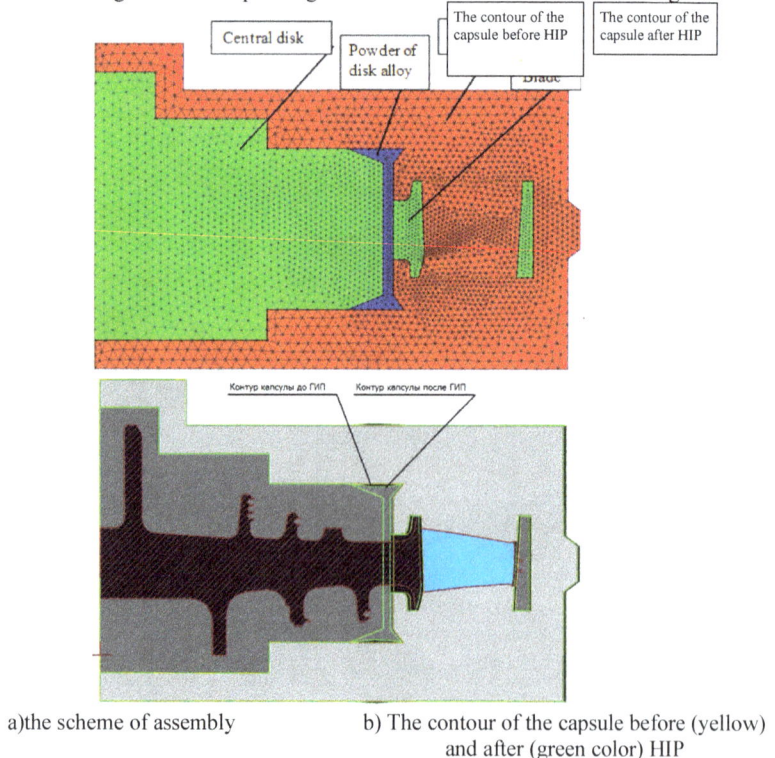

a)the scheme of assembly                    b) The contour of the capsule before (yellow)
                                                         and after (green color) HIP

*Fig. 6 Simulation results for the shrinkage of the HIP tooling*

The deformation of the capsule assembly during the process of hot isostatic pressing using FEA is calculated to determine the initial position of the blades with shrouds, to enable the mounting preload after the HIP and heat treatment. For the developed structure of the capsule assembly, the value of the initial gap between the central disk and the annular assembly of the blades and inserts was determined to be 5 mm. This gap is necessary for reliable filling of the powder material and its sustainable compaction. The initial density of the powder in the annular gap after vibration filling was 65%.

Analysis of the calculation results showed that the shrinkage of the powder is stable and occurs mainly in the radial direction. The value of the interference fit between the blade shrouds can be controlled by adjusting the magnitude of the gap.

The technological process of manufacturing of blisks with shrouded blades of the cast superalloy and disc from powder alloy is the following:

- Casting of the blades.
- Manufacturing of capsules for the central part of the disk and HIP of the central disc to minimize shrinkage at the final shaping of blisk.
- Manufacture of capsule assembly to connect the disc part and blades and the inserts.
- The assembly in the capsule tooling with blades and the disc part.
- Conduct of the HIP.
- Heat treatment and removal of the capsule and inserts, machining.

## Summary

As a result of this work the technological process of manufacturing BB with shrouded blades by hot isostatic pressing method has been suggested and developed.

The capsules for forming the central part of disk and for assembling disk and blades were developed.

FEM Modeling of capsule and powder deformation during hot isostatic pressing process was carried out.

The elements of the tooling and technological process were designed.

## References

[1] V. C. Reelnikov, A. N. Afanaciev-Hodeekin, I. A. Galushka, Soldering technology for type design "blisk" of dissimilar alloys, TRUDY VIAM, Publisher: all-Russian scientific research Institute of aviation materials (Moscow) ISSN: 2307-6046e

[2] Information on https://www.theengineer.co.uk/supplier-network/product/delcam-and-technicut-partner-to-reduce-time-and-cost-of-blisk-machining/

[3] Information on http://www.indiandefensenews.in/2016/07/idn-take-tech-scan-brief-review-of.html

[4] Information on https://www.aero-mag.com/aml-plaudits-rolls-royce-blisk-manufacture/

[5] W. Gang, L. Li-Hui, H. Xi-Na, L. Jia-Jie. The shielding effect of capsule in the process of Hot Isostatic Pressing (HIP), 13th International Bhurban Conf. on Applied Sciences and Technology (IBCAST) Islamabad, Pakistan 2016.

[6] L. Redouani, S. Boudrahem Hot isostatic pressing process simulation: application to metal powders: Laboratory of Technology of Materials and Engineering of the Processes, Faculty of Technology, University of Bejaia, Bejaia 06000, Algeria, Published on the web 31 May 2012.

[7] L. A. Magerramova, T.P. Zaharova, M.V. Gromov. Optimal Design of Bimetallic Components Manufactured by HIP from Powder and Cast Ni-base Super alloys, Proc. HIP99, Beijing, pp. 229-238.

[8] T.E. Strangman, Integral turbine composed of a cast single crystal blade ring diffusion bonded to a high strength disk, 2006, European Patent Application EP1728971.

[9] L.A. Magerramova ,The project development of HIPed Bimetallic Blisk with Uncooled and Cooled Blades for Engine Gas Turbine of Difference Using, Proc. Int. Conf. HIP14, id: 140605-1035, Stockholm.

[10] L.A. Magerramova, E.P. Kratt, V.V. Yacinsky, A method of manufacturing integral blisks with uncooled rotor blades for a gas turbine engine and integrated blisk, 2012, Russian patent #2457177.

[11] L.A. Magerramova, E.P. Kratt, V.V. Yacinsky, A method of manufacturing integral blisks with cooled blades, integrated blisk and cooled blade for gas turbine engine, 2013Russian patent #2478796.

[12] L.A. Magerramova, E.P. Kratt, A method of manufacturing an integrated turbine blisks of various metal alloys for gas turbine engine, 2016, Russian patent #2579558.

[13] L.A. Magerramova, S.S. Bakulin, The method of positioning the blades in the manufacture of integral blisks of gas turbine engine turbine engine, 2016, Russian patent #2595331.

[14] L.A. Magerramova, R.Z. Nigmatullin, B.E. Vasilyev, V.S. Kinzburskiy, Design of a bimetallic blisk turbine for a gas turbine engine and its production using powder metallurgy methods. Proc. of ASME Turbo Expo 2017 Turbomachinery Technical Conference & Exposition GT2017 June 27-29, 2017, Charlotte, NC, USA GT2017-63560.

Hot Isostatic Pressing – HIP'17
Materials Research Proceedings **10** (2019) 47-57

Materials Research Forum LLC
doi: http://dx.doi.org/10.21741/9781644900031-7

# Large-Scale and Industrialized HIP Equipment for the Densification of Additive Manufactured Parts

Hongxia Chen[*], Deming Zhang and Qing Ye

CHINA IRON& STEEL RESEARCH INSTITUTE GROUP (CISRI), No. 76 Xueyuan Nanlu, Haidian District, Beijing 100081, P. R. China

iptec@163.com

**Keywords:** Hot Isostatic Pressing (HIP), Additive Manufacturing, Temperature Uniformity

**Abstract.** Additive manufacturing technology has significant advantages in fabricating parts with complex shape, but the internal defects, such as residual stress, pores and microcracks, would result in critical problems under certain circumstances. To meet the requirement of HIP treatment on additive manufactured parts, we studied the thermodynamic behavior of the gas medium under high temperature and high pressure conditions, explored the deformation discipline of the thin-walled parts and the boundary conditions of controlling deformation, and optimized the process of eliminating residual stress. Based on the above work, a series of HIP equipment were specially designed for the treatment on additive manufactured parts, which could provide solid support for the development of additive manufacturing technology.

## 1. Advantages of additive manufacturing technology

Additive manufacturing is an advanced technology widely developed in the world. Based on the material, it contains rapid forming of plastic, wax, ceramic and metal. Among those, rapid forming of wax-based material combines additive manufacturing and casting technology, which has been widely used for the industrial production of castings. Besides, additive manufacturing of metallic material is the most impressive, which provides universal process for the direct fabricating of key parts with complex shape. With no need of mold and the short manufacturing cycle, it has become the best process for preliminary examination and small-batch production.

Recently, additive manufacturing of metallic material mainly focus on super alloy, ultra-high strength steel, titanium alloy and aluminum alloy. Due to the high cost, the application areas are limited in aerospace, military and biomedical industries. In the future, with the development of additive manufacturing and reducing of cost, it will play an important role in more fields.

## 2. Defects of additive manufactured parts and improvement

### 2.1 Analysis on defects of additive manufactured parts

Compared with traditional technology, additive manufacturing has several advantages. However, due to the uniqueness of the forming process, internal defects tend to appear easily [1]: Types of internal defects result from 1) Residual stress: Thermal strain and residual stress generate due to the high temperature gradient; 2) Spherulization effect: When the laser or electron beam is irradiated, metal powders are partially melted to form a molten pool. Under a certain force, the melt tends to be spherical, resulting in poor surface quality, low density and emergence of pores; 3) Cracks: In the forming process, metal powders undergo rapid heating and cooling. There is no enough liquid metal supplementation during solidification, and the solidification part is bound by the cold substrate, resulting crack. 4) Pore formation: Pores may generate from residual gas during rapid solidification, reaction of carbon and oxygen in the melt, reduction of the metal oxide by carbon, volatile of solid material, evaporate of moisture and coagulation shrinkage of sintered layer. Although defects could be reduced by optimization of the process parameters, producing large defect-free parts with complex shape through additive manufacturing is still a great challenge.

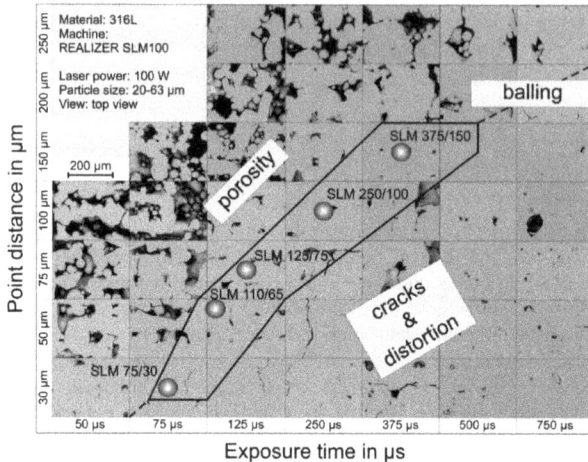

Fig. 1 Effects of SLM parameters on the micro-defects[1]

Compared with traditional technology, fabricated additive manufactured parts show lower mechanical performance because of defects. Besides, due to the manufacturing orientation, the performance shows anisotropic character. Rottger [1] studied the mechanical properties of 316L stainless steel prepared by selective laser melting (SLM) process in different forming directions. It was found that the tensile and yield strength of the specimens built in the horizontal direction are usually higher than in the vertical direction (Tensile specimens were built-up horizontal as well as vertical to the building platform). The reason is that the connection of a molten pool to the neighboring areas in the same slice layer is better than the connection to the underlying slice layer. Performance degradation and anisotropic distribution caused by defects will greatly affect the application of key parts, especially those with complex shape and thin walls. Therefore, after additive manufacturing, further eliminating the defects through HIP  improving properties becomes a more critical process.

### 2.2 Research progress of HIP post-treatment on additive manufactured parts

Hot isostatic pressing is the best process to improve the performance of additive manufactured parts. Many researchers studied the effect of HIP treatment on the defects, microstructure and mechanical properties of additive manufactured 316L stainless steel [2], titanium alloy Ti-6Al-4V [3-7] and super alloy [8, 9, 10]. Results show that the effect of HIP on the elimination of residual pores was related to the initial state of powder material and the atmosphere during additive manufacturing process. For the pores formed inside the original powder particles during atomization process, residual argon was trapped in the pores and could not disperse to the surface at high temperature because of large atom size. Therefore, under the high pressure of HIP, the size of the pore was gradually reduced. While the internal pressure was equal to the HIP working pressure, the pore size reached the limit. Furthermore, if under subsequent high-temperature heat treatment, these types of pores are likely to re-grow. A related study found and confirmed this result (seen in Fig. 2[6]). However, for the pores and microcracks formed by incomplete fusion of powders during additive manufacturing process, the effect of HIP on the elimination of residual pores was related to the process atmosphere. SLM process is carried out under argon atmosphere, so part of argon will be blocked in the residual pores or microcracks, resulting in incomplete closing of defects during HIP treatment. However, the argon pressure during SLM process is much lower than that during atomization, the closing effect of HIP on pores and microcracks from additive manufacturing will be better than that formed from atomization. Electron beam melting (EBM) process works under vacuum, and thus the effect of HIP on the elimination of residual pores and microcracks could achieve the best results because there is no residual gas inside the defects.

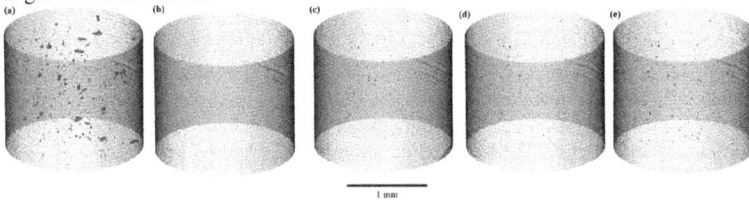

*Fig. 2 3D visualisation of the porosity (red) imaged by CT scans of the same cylindrical sample (build direction vertical) (a) as-built; (b) following HIPing; (c) 10min at 1035 °C; (d) 10 h at 1035 °C; and (e) 10 min at 1200 °C. [6]*

Despite the fact that the process atmosphere affects the densification effect of HIP on additive manufactured parts, HIP process still contributes to the amount and size reduction of residual pores and microcracks, even for SLM parts manufactured under argon atmosphere. Therefore, HIP has an excellent effect of eliminating the residual pores and microcracks in additive manufactured parts and improving the mechanical properties (seen in Fig. 3 and Fig. 4 [4]), ensuring applications under various conditions.

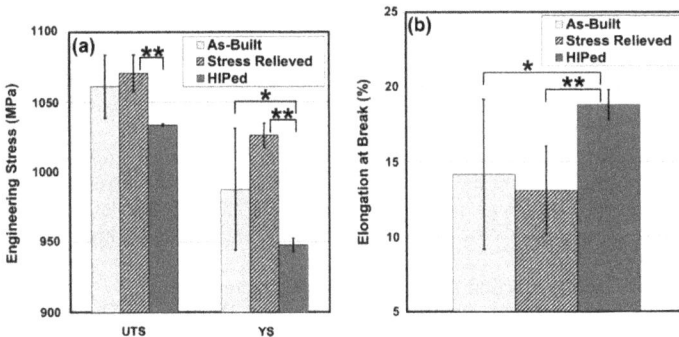

*Fig. 3 Tensile properties for three conditions: As-Built, stress relieved and HIPed.[4]*

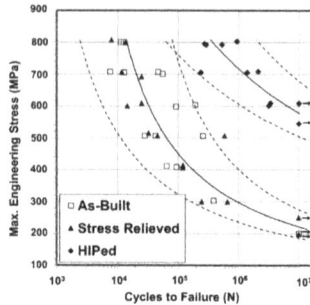

*Fig. 4 S–N curve fatigue results for all three conditions: As-built, stress relieved and HIPed. [4]*

Treatment of residual stress is different from that of pores and microcracks. On the one hand, the residual stress could be reduced by decreasing the temperature gradient through preheating of the powder bed (as for EBM process). On the other hand, the residual stress could be relieved by subsequent annealing process (The stress relief parameters for additive manufactured Ti-6Al-4V: 650 °C, 5h). Besides, the stress relief process can be carried out in a conventional annealing furnace and there is no need to apply such treatment in HIP equipment. Therefore, for additive manufactured parts, stress relief annealing will be firstly carried out, and then HIP treatment will be applied to reduce or eliminate pores and microcracks. Finally, heat treatment is utilized to achieve the required mechanical properties.

## 3. Characteristics of CISRI-HIP for industrial application of additive manufactured parts

### 3.1 Characteristics and analysis of CISRI-HIP

### 3.1.1 Characteristics of CISRI-HIP

With the development of additive manufacturing technology, key parts tend to be large and complex. In response to this trend and the urgent need for hot isostatic pressing to accommodate these complex parts, CISRI fully upgraded the existing hot isostatic pressing equipment and developed specialized ultra-large HIP equipment for large-scale production of additive manufactured parts. The specific characteristics of this equipment are as follows:

1) CISRI has developed extra large HIP with diameter > 2100 mm and height > 4500 mm, to meet the HIP post-treatment demands of additive manufactured parts with larger size, more complex shape and more internal defects.

2) CISRI-HIP has the function of preventing the deformation of the additive manufactured parts from happening again, to meet the demands of stability for additive manufactured parts during charging and under high pressure condition. For this, CISRI developed deformation prevention devices to improve the stability during charging and under high-pressure air flow condition.

3) CISRI-HIP has the function of ensuring the temperature uniformity for single large-size parts or high-volume products with small size during HIP process. Based on the existing technology on controlling temperature uniformity, CISRI has upgraded to further improve the uniformity, to ensure the uniform temperature in every part of a single large component or each workpiece of mass production.

CISRI-HIP has the function of uniform temperature in the large hot zone under high-temperature and high-pressure condition, to ensure that the large additive manufactured parts have a uniform temperature during HIP densification and there should not exist any thermal stress or thermal deformation due to the deviation of temperature under high pressure.

4) CISRI-HIP has the function that the high-pressure air flow sweeps various parts of the large parts uniformly in the large hot zone under high temperature and high pressure condition, to ensure that there should not exist any thermal deformation or thermal stress due to the high-speed gas flow disturbance of any direction.

Focusing on the characteristics of CISRI-HIP, this paper discusses additional features in detail.

### 3.1.2 Control, calibration and feedback control of temperature accuracy [11]
Choose PID control with precision control accuracy.

#### 1) Control Law of Proportional Integral Derivative (PID)
The transfer function of PID control is

$$G_c(s) = K_p + \frac{1}{T_i S} + T_d \bullet s = \frac{T_i T_d s^2 + K_p T_i s + 1}{T_i s} = \frac{(\frac{1}{T_1}s+1)(\frac{1}{T_2}s+1)}{T_2 s} . \tag{1}$$

When Kp = 1, $G_c(j\omega) = 1 + \frac{1}{jT_i\omega} + jT_d \bullet \omega$, the corresponding Bode diagram is shown in

Fig. 5.

As can be seen from Fig. 5, when Ti> Td, PID control plays an integral role in the low frequency to improve the steady-state characteristics of the system, and it plays a differential role in the middle stage to improve the dynamic characteristics of the system. From Eq. 1, we can see that PID correction adds one zero pole and two negative real zero points. The zero pole increases the degree of system indifference by one level and improves the steady-state accuracy. Through proper regulation, the two negative real zero points could improve the dynamic performance of the system, so PID control has been widely used in engineering.

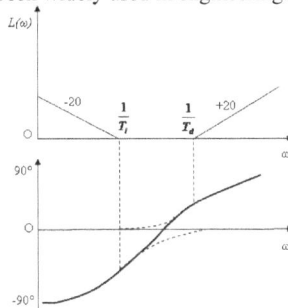

Fig. 5 Bode diagram of PID control

#### 2) Calibration and feedback control
Phase lag lead correction

The purpose of lead correction is to improve the relative stability of the system and quick response. The main function of lag correction is to improve the low-frequency gain and the steady-state characteristics of the system without affecting the transient performance. Lag lead correction can improve the transient and steady-state characteristics of the system at the same time. The essence of this kind of correction is to make full use of the respective characteristics of lag and lead correction, improving the transient characteristics by using its leading part and the steady-state characteristics by utilizing the lagging part.

### 3) Feedforward and feedback compound control

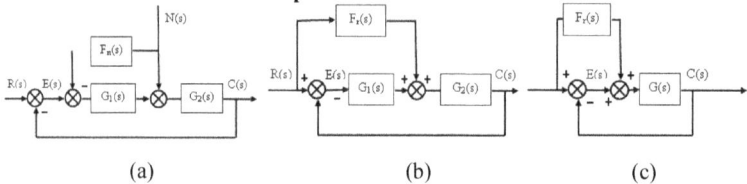

(a)                 (b)                (c)

*Fig. 6 (a): Disturbance feedforward compound control; (b): Input feedforward compound control; (c): Compound control system.*

#### a Compound control according to disturbance compensation

Fig. 6 shows a block diagram of a compound control system according to disturbance feedforward compensation. $F_n(s)$ is the transfer function of the feedforward compensation device. If R (s) = 0, then

$$C(s) = [N(s) - (C(s) + F_n(s)N(s)G_1(s)]G_2(s). \tag{2}$$

That is $C(s) = \dfrac{1 - F_n(s)G_1(s)]G_2 s}{1 + G_1(s)G_2(s)} N(s). \tag{3}$

If choosing the transfer function of the feedforward device as

$$F_n(s) = \frac{1}{G_1(s)}. \tag{4}$$

Then C(s)=0  E(s)=-C(s)=0

This shows that the output response C(s) is completely undisturbed by the influence of disturbance N(s), so the transient and steady-state error E(s) of the disturbed system are all zero. Eq. 4 is called the full compensation condition for disturbance error.

#### b Compound control according to input compensation

Figure 6b shows a compound control system according to input compensation. The system has two forward channels: one is a feedforward channel formed by Fr(s)G2(s), which is controlled by open loop and the other is the main control channel formed by G1(s)G2(s), which is controlled by closed loop. According to the principle of superposition of linear systems:

$$C(s) = [(R(s) - C(s)G_1(s) + R(s)F_r(s)]G_2(s). \tag{5}$$

or

$$C(s) = \frac{F_r(s)G_2(s) + G_1(s)G_2(s)}{1 + G_1(s)G_2(s)} R(s). \tag{6}$$

If choosing the feedforward compensation device as

$$F_r(s) = \frac{1}{G_2(s)}. \tag{7}$$

The Eq. 6 becomes

$C(s)=R(s)$   $E(s)=R(s)-C(s)=0$

This Equation shows that the system output fully reproduce the input after feedforward compensation, so the transient and steady-state errors are zero. Eq. 7 is called the error full compensation of the input signal.

In order to analyze the error and the stability of the input feedforward compound control system, the concept of equivalent transfer function is introduced. For the compound control system shown in Fig. 7, the equivalent closed-loop transfer function can be obtained from Eq. 6.

$$\Phi(s) = \frac{C(s)}{R(s)} = \frac{G_1(s)G_2(s) + F_r(s)G_2(s)}{1 + G_1(s)G_2(s)}. \tag{8}$$

Define the equivalent open-loop transfer function as

$$G_K(s) = \frac{C(s)}{E(s)} = \frac{\Phi(s)}{1 - \Phi(s)}. \tag{9}$$

Obviously, the above two equations are only applied to the unit feedback compound control system. The error transfer function is

$$\Phi_e(s) = \frac{E(s)}{R(s)} = 1 - \Phi(s). \tag{10}$$

For the convenience of discussion, let $G1(s) = 1$ in Fig. 6b, so the compound control system is obtained and shown in Fig. 6c. The following is a discussion of the partial compensation of $Fr(s) \neq 1/G(s)$.

Let

$$G(s) = \frac{K_v}{s(a_n s^{n-1} + a_{n-1} s^{n-2} + \cdots + a_2 s + a_1)}. \tag{11}$$

The closed-loop transfer function is

$$\Phi(s) = \frac{G(s)}{1 + G(s)} = \frac{K_v}{s(a_n s^{n-1} + a_{n-1} s^{n-2} + \cdots + a_2 s + a_1) + K_v}. \tag{12}$$

This is the closed-loop transfer function of the original feedback system without feedforward channel. Obviously, this is a type I system, with constant error and infinite acceleration error.

(1) Take the first derivative of r(t) as the feedforward control signal, that is, $\lambda_1$ is a constant, which indicates the strength of the feedforward signal. The equivalent open-loop transfer function can be obtained

$$G_K(s) = \frac{1 - \Phi_e(s)}{\Phi_e(s)} = \frac{a_1 s + K_v}{s^2(a_n s^{n-2} + \cdots a_2)} . \tag{13}$$

Obviously, the type of system is increased from type I (v=1) to type II (v=2).

(2) Taking the linear combination of the first derivative and the second derivative of r(t) as the feedforward control signal, that is $F_r(s) = \lambda_1 s + \lambda_2 s^2$. The equivalent open-loop transfer function can be obtained.

$$G_K(s) = \frac{a_2 s^2 + a_1 s + K_v}{s^3(a_n s^{n-3} + a_{n-1} s^{n-2} + \cdots + a_3)} . \tag{14}$$

Obviously, the type of system is increased from type I to type III.

In summary, we can draw the following conclusions:

When the input feedforward signal is introduced into the feedback control system, the type of the system can be increased from type I to type II and the speed error is eliminated from the system if the feedforward channel transfer function is taken as $F_r(s) = \lambda_1 s = \dfrac{a_1}{K_v} s$. If it is taken as $F_r(s) = \lambda_1 s + \lambda_2 s^2 = (\dfrac{a_1}{K_v} s + \dfrac{a_2}{K_v} s^2)$, the system type can be increased from type I to type III, eliminating the acceleration error and thus greatly improving the ability and accuracy of the system to reproduce the input signal. Like the disturbance feedforward, the introduction of input feedforward does not change the stability of the original system, so it solves the contradiction between improving control precision and ensuring system stability.

### 3.1.3 Energy distribution of high-pressure air flow
The flow field is two-dimensional steady-state, that is

$$\frac{\partial(\rho u)}{\partial x} + \frac{\partial(\rho v)}{\partial y} = 0 . \qquad \text{Momentum equation} \tag{15}$$

$$\rho u \frac{\partial u}{\partial x} + \rho v \frac{\partial u}{\partial y} = \frac{-d\rho}{dx} - \rho g + \frac{\partial(u \frac{\partial u}{\partial y})}{\partial y} . \qquad \text{Energy equation} \tag{16}$$

$$\frac{d\rho}{dx} = -\rho_\infty g . \qquad \text{Pressure gradient term} \tag{17}$$

Combining the body force and pressure gradient term, the momentum equation becomes

$$\rho u \frac{\partial u}{\partial x} + \rho v \frac{\partial u}{\partial y} = g(\rho_\infty - \rho) + \frac{\partial(u \frac{\partial u}{\partial y})}{\partial y}. \tag{18}$$

$$\rho u \frac{\partial u}{\partial x} + \rho v \frac{\partial u}{\partial y} = g(\rho_\infty - \rho) + \frac{\partial(u \frac{\partial u}{\partial y})}{\partial y}. \tag{19}$$

From this, we can see that the thermal field is in a forced convection condition with the increase of temperature and pressure under high temperature and high pressure condition. The gas density increases, the pressure gradient increases, and the gas velocity increases, that is, the kinetic energy increases. So the impact of the air flow on the workpiece is also increasing. Therefore, it is necessary to design the airflow direction, flow force and energy distribution reasonably, that is, uniform flow force and rational energy distribution.

## 3.2 Serialized and specialized HIP of CISRI for additive manufactured parts

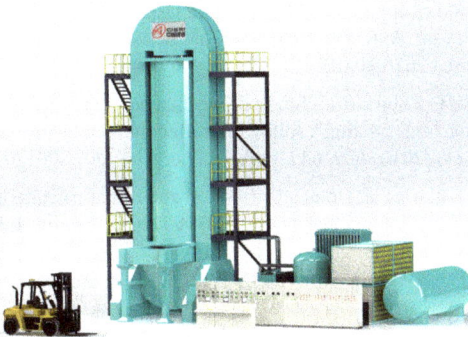

*Fig. 7 CISRI-HIP for additive manufactured parts*

Based on conventional HIP equipment, CISRI developed specialized ultra-large HIP equipment for large-scale production of additive manufactured parts (shown in Fig. 7). According to the working pressure and temperature, Table 1 summarises the specific features of the serialized HIP machines.

*Table 1 CISRI serialized HIP for additive manufactured parts*

| Pattern | Highest pressure [MPa] | Highest temperature [°C] | Maximum hot zone [mm] | Working media |
|---------|------------------------|--------------------------|-----------------------|---------------|
| HIP1250 | 200 | 1400 | Φ1250×2500 | Ar≥99.99% |
| HIP1600 | 200 | 1400 | Φ1600×3500 | Ar≥99.99% |
| HIP1800 | 150 | 1400 | Φ1800×4000 | Ar≥99.99% |
| HIP2500 | 120 | 650 | Φ2500×4500 | Ar≥99.99% |
| HIP3500 | 100 | 650 | Φ3500×(5000~6000) | Ar≥99.99% |

**Summary**
1) HIP post-treatment is the best process to eliminate the internal defects and improve the mechanical properties of additive manufactured parts.
2) CISRI-HIP has the function of preventing the deformation of additive manufactured parts. 3) CISRI-HIP has the function of uniform temperature in the large hot zone under high-temperature and high-pressure condition, to ensure that the large additive manufactured parts have a uniform temperature during HIP densification and there should not exist any thermal stress or thermal deformation due to the deviation of temperature under high pressure.
4) CISRI-HIP has the function that the high-pressure air flow sweeps various parts of the large parts uniformly in the large hot zone under high temperature and high pressure condition, to ensure that there should not exist any thermal deformation or thermal stress due to the high-speed air flow disturbance of any direction.
5) Additive manufactured parts tend to be large, complex and integrate. For the large-scale HIP treatment, CISRI developed ultra-large and specialized HIP equipment for additive manufactured parts.

**References**

[1] A. Rottger, K. Geenen, M. Windmann, et al. Comparison of microstructure and mechanical properties of 316 L austenitic steel processed by selective laser melting with hot-isostatic pressed and cast material, Mat. Sci. Eng. A, 678 (2016) 365-376. https://doi.org/10.1016/j.msea.2016.10.012

[2] I. Tolosa, F. Garciandia, F. Zubiri, et al. Study of mechanical properties of AISI 316 stainless steel processed by "selective laser melting", following different manufacturing strategies, Int. J. Adv. Manuf. Technol., 51 (5) (2010) 639–647. https://doi.org/10.1007/s00170-010-2631-5

[3] M. Seifi, A. Salem, D. Satko, et al. Defect distribution and microstructure heterogeneity effects on fracture resistance and fatigue behavior of EBM Ti–6Al–4V, Int. J. Fatigue, 94 (2017) 263-287. https://doi.org/10.1016/j.ijfatigue.2016.06.001

[4] N. Hrabe, T. Gnäupel-Herold, T. Quinn. Fatigue properties of a titanium alloy (Ti–6Al–4V) fabricated via electron beam melting (EBM): Effects of internal defects and residual stress, Int. J. Fatigue, 94 (2017) 202-210. https://doi.org/10.1016/j.ijfatigue.2016.04.022

[5] B. V. Hooreweder, Y. Apers, K. Lietaert, et al. Improving the fatigue performance of porous metallic biomaterials produced by Selective Laser Melting, Acta Biomater. 47 (2017) 193-202. https://doi.org/10.1016/j.actbio.2016.10.005

[6] S. Tammas-Williams, P.J.Withers, I. Todd, P.B. Prangnell. Porosity regrowth during heat treatment of hot isostatically pressed additively manufactured titanium components [J]. Scripta Mater., 122 (2016) 72-76. https://doi.org/10.1016/j.scriptamat.2016.05.002

[7] M. W. Wu, P. H. Lai. The positive effect of hot isostatic pressing on improving the anisotropies of bending and impact properties in selective laser melted Ti-6Al-4V alloy, Mat. Sci. Eng. A, 658 (2016) 429-438. https://doi.org/10.1016/j.msea.2016.02.023

[8] B. Ruttert, M. Ramsperger, L. Mujica Roncery, et al. Impact of hot isostatic pressing on microstructures of CMSX-4 Ni-base superalloy fabricated by selective electron beam melting, Mater. Design 110 (2016) 720-727. https://doi.org/10.1016/j.matdes.2016.08.041

[9] M. M. Kirka, F. Medina, R. Dehoff, et al. Mechanical behavior of post-processed Inconel 718 manufactured through the electron beam melting process, Mat. Sci. Eng. A, 680 (2017) 338-346. https://doi.org/10.1016/j.msea.2016.10.069

[10] M. E. Aydinöz, F. Brenne, M. Schaper, et al. On the microstructural and mechanical properties of post-treated additively manufactured Inconel 718 superalloy under quasi-static and cyclic loading, Mat. Sci. Eng. A, 669 (2016) 246-258. https://doi.org/10.1016/j.msea.2016.05.089

[11] J. G. Zhang. Principle of Automatic Control, Harbin Institute of Technology Press, Harbin, 2003, pp. 129-158. (in Chinese)

# HIP Process of a Valve Body to Near-Net-Shape using Grade 91 Powder

Toyohito Shiokawa[1,a,*], Hiroshi Urakawa[2,b], Mitsuo Okuwaki[1,c], Yuto Nagamachi[3,d]

[1]970 Shimoise, Hayashida-cho, Himeji-shi, Hyogo-ken, 679-4233, JAPAN

[2]250, Ryusen-cho, Aioi City, Hyogo, JAPAN

[3]713 Shake-aza, Narihira, Ebina-shi, Kanagawa-ken, 243-0424, JAPAN

[a]tshiokawa@kinzoku.co.jp, [b]h.urakawa@shimoda-flg.co.jp, [c]okuwaki@kinzoku.co.jp, [d]ynagamachi@kinzoku.co.jp

**Keywords:** Near-Net-Shape, Grade 91, HIP

**Abstract.** Materials used for steam piping of power plants are exposed to high temperatures and high pressures over long periods of time. As a consequence, forged Grade 91 alloy steel is commonly chosen to meet these demands. However, complicated structures such as a valve body are often machined from large forged blocks resulting in long machining time and the material weight being heavy. Therefore, by manufacturing a valve body with near net technology, both the time and material weight can be reduced. This paper will present 1) a survey of the dimensions of a valve body HIPed to a near net shape, 2) an investigation of the mechanical properties of Grade 91 powder, 3) a comparison of the structure of a HIPed product and a forged product, 4) The machining time and material weight of a near net shaped (NNS) product by HIP compared to a forged product. This paper will illustrate that the NNS product was able to reduce the machining time by 30% and the material weight by 40% less than when machining from a forged product.

## Introduction

Grade 91, which is a high chrome steel, has excellent high temperature strength and is therefore used in steam piping of power plants which are exposed to high temperatures and high pressures for long periods of time. Due to this environment heavy loads are placed on the piping. Most of these parts are currently manufactured by forging, and large parts such as a valve body are assembled and welded using multiple small parts, or are machined from one large block. However, the seam areas, such as the welded portion, have a lower strength to the base metal. In order to prolong the life of the steam piping, there is a desire to reduce the number of welds and fabricate an integrated structure by machining. Since the integrated structure has a complex shape, it must be machined from a large block, machining time and material weight increases, which is expensive. Components produced with PM/HIP can be produced very near final shape, resulting in reduced component weight, reduced machining, reduced energy, and reduced waste from post-process machining. [1] This paper will present 1) a survey of the dimensions of a valve body HIPed to a near net shape, 2) an investigation of the mechanical properties of Grade 91 powder, 3) a comparison of the structure of a HIPed product and a forged product, 4) The machining time and material weight of a NNS product by HIP compared to a forged product.

## Method

Fig. 1 is the post HIP target shape dimensions. Fig. 2 is the capsule's dimensions based on the post HIP target shape.The capsule designed had an extra 5mm excess thickness on each side and we assumed a packing density of 67%. The Circles in fig. 2 indicate an additional 20mm thickness to aid welding. The grey block in the centre indicates where we placed a mild steel block which also aided the welding process of the capsule. Since the capsule rigidity (thickness) greatly affects the shape of the HIPed products and the deformation during densification becomes more isotropic when the capsule materials are thin and the same thicknesses are used [2] the capsule material was constructed from mild steel with a thickness of 3.2mm.

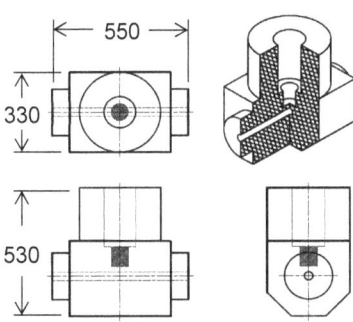

Fig. 1 Post HIP target shape                  Fig. 2 Capsule shape

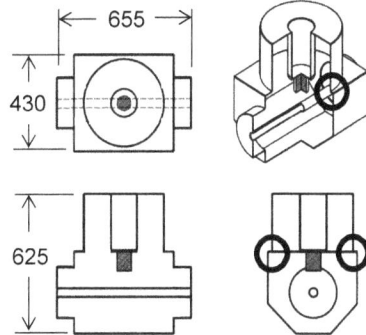

The Powder used was Grade 91; the chemical composition is shown in table 1. The Actual packing density that was reached was 67.4%. The capsule was then HIPed at 1200 degrees Celsius and 118 megapascals. After HIP, the capsule dimensions were measured for a survey of the near net shape valve body by HIP.

| Table 1. Chemical composition | | | | | | | | |
|---|---|---|---|---|---|---|---|---|
| C | Mn | P | S | Si | Cr | Mo | Ni | V |
| 0.10 | 0.39 | 0.011 | 0.005 | 0.40 | 8.77 | 0.90 | 0.17 | 0.25 |
| Cd | Cu | Al | Ti | Sn | W | N | As | Zr |
| 0.08 | 0.09 | 0.00 | 0.003 | 0.003 | 0.00 | 0.050 | 0.001 | 0.002 |
| Sb | Pb | O | | N/Al > 24 | | | | |
| <0.002 | 0.00006 | 110ppm | | Sb + Sn + As + Pb <0.0061% | | | | |

After the measurements were taken the capsule was removed from the HIPed product by machining and heat treated according to ASTM A989 and the following investigation was conducted.
· An investigation of the mechanical properties of NNS Grade 91
· A comparison of the structure of a HIPed product and a forged product
· A comparison of the machining times and material weights of both products.

## Results and Discussion
### A survey of the dimensions of a near net shape valve body by HIP

A comparison of the dimensions before and after HIP and the shrinkage ratio of the capsule were calculated. Shrinkage ratio = {(|dimensions before HIP - dimensions after HIP|)/dimensions before HIP processing} x 100% were calculated. The target shape and the actual shape were compared and the difference was investigated. The difference between the target shape and the post-HIP shape was calculated as {(| target dimension - post-HIP dimension minus capsule wall thickness |) / target dimension} x 100%. Fig. 3 and table 2 show the measurement positions and results. In table 2 the dimensions of E show the figures with the 20mm 'extra welding support' *removed. T*he bottom surface was under the target shape of 6.7mm. The upper side surface and the intermediate surface were both around 20mm.The shrinkage rate of the capsule was 4.3 to 11.8%. The shrinkage ratios of C and D were as low as 4.3 and 5.1%. This was due to the small diameter which helped capsule stiffness and decreased the shrinkage rate. The outer sides were 10.9 - 11.8% and they contracted almost equally. The difference from the overall target size after HIP is now only 3.2 to 5.5 percent. This was the first time to manufacture this kind of valve body. Our next goal is to keep the difference between the shape after HIP and the target shape within 2%.

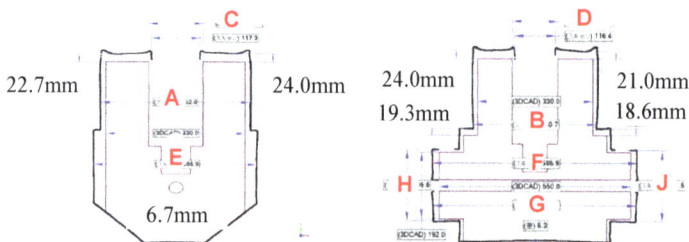

*Fig. 3 The measurement positions and results.*

| Table 2 Dimension measurements results | A | B | C | D | E | F | G | H | J |
|---|---|---|---|---|---|---|---|---|---|
| Before HIP [ mm ] | 396.5 | 396.5 | 122.6 | 122.6 | 396.0 | 658.9 | 659.0 | 231.7 | 231.4 |
| After HIP [ mm ] | 352.0 | 350.7 | 117.3 | 116.4 | 350.5 | 586.9 | 581.3 | 205.6 | 204.5 |
| Shrinkage ratio [ % ] | 11.2 | 11.6 | 4.3 | 5.1 | 11.5 | 10.9 | 11.8 | 11.3 | 11.6 |
| Goal shape [ mm ] | 330 | 330 | 130 | 130 | 330 | 550 | 550 | 192 | 192 |
| Error [ % ] | 4.7 | 4.3 | 4.8 | 5.5 | 4.3 | 5.5 | 4.5 | 3.7 | 3.2 |

## An investigation of the mechanical properties of NNS Grade 91

The Capsule was removed from the HIPed product by machining and heat treated according to ASTM A989. After heat treatment, the HIPed body was cut and we performed a tensile test at room temperature, an elevated temperature tensile test, and a Brinell hardness test. The tensile test samples at room temperature and the elevated temperature tensile test samples were taken from each of the places indicated in fig. 4. These samples were taken from two directions, one in the axial direction and the other in the circumferential direction.

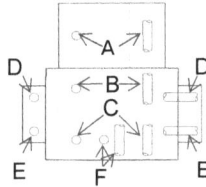

*Fig. 4 Position of both Tensile Tests Samples*

Table 3 shows the tensile test results at room temperature. The Tensile test was conducted to investigate 0.2% proof stress, tensile strength, elongation value, and reduction of area value. A to F are the average values of the tensile test results in the axial direction and circumferential direction. There is almost no difference between the tensile test results in the axial direction and the circumferential direction. The average displayed is the average for all of the samples. All tests were higher than the standard values of ASTM A989.

| Table 3. The tensile test results at room temperature | | | | | | | | |
|---|---|---|---|---|---|---|---|---|
| | A | B | C | D | E | F | Average | ASTM A 989 |
| 0.2% proof stress [ MPa ] | 475.0 | 474.3 | 474.5 | 472.5 | 474.5 | 474.3 | 474.2 | 415 |
| Tensile strength [ MPa ] | 648.5 | 649.3 | 645.8 | 645.5 | 646.5 | 644.8 | 646.7 | 585 |
| Elongation [ % ] | 29.2 | 29.8 | 28.5 | 28.3 | 28.5 | 28.7 | 28.8 | 20 |
| Reduction of area [ % ] | 71.1 | 71.2 | 70.2 | 71.2 | 70.8 | 70.7 | 70.9 | 40 |

Fig. 5 shows the results from the elevated temperature tensile test. These tests were also performed at intervals of 100 degrees Celsius at a range of 100 degrees Celsius up to 700 degrees Celsius. From 400 degrees Celsius the 0.2 percent proof stress and tensile strength tests results were weaker. However, the results for the elongation and reduction of area tests were more ductile.

Hot Isostatic Pressing – HIP'17                                    Materials Research Forum LLC
Materials Research Proceedings **10** (2019) 58-64          doi: http://dx.doi.org/10.21741/9781644900031-8

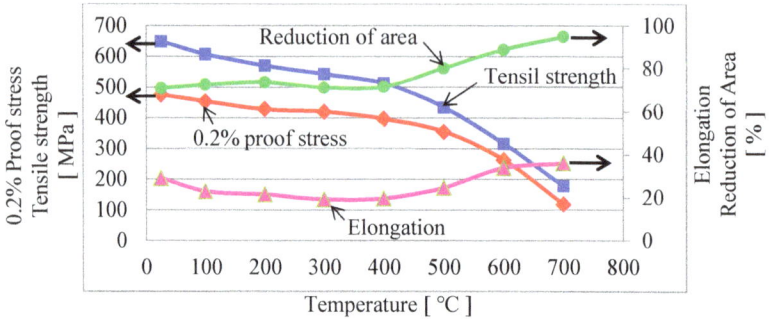

*Fig. 5 The results from the elevated temperature tensile*

Fig. 6 is the Brinell hardness measurement results. The hardness of the HIPed body was HB 197 to 205, and there was almost no difference between the surface layer and the interior. All the mechanical properties satisfied ASTM A 989 standard. Currently 3,000 hours and 10,000 hours creep tests are also conducted.

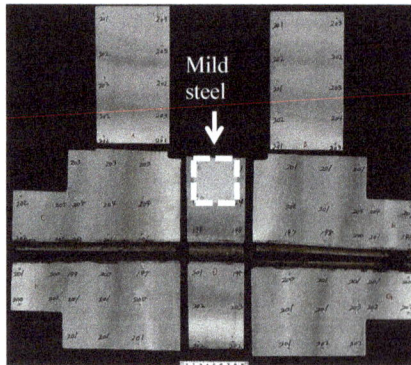

*Fig. 6 The Brinell hardness measurement results*

**A comparison of the structure of a HIPed product and a forged product**

Microstructure observation of the HIPed product after the heat treatment was carried out to investigate the difference between the structure of the HIPed product and the forged product. Nonmetallic inclusions of HIPed products were also measured. Fig. 7 shows the microstructure of grade 91. On the left is the HIPed product and on the right is the forged product. The forged product has a distinct visible fiber flow. However, in contrast the HIPed product's fiber flow is not visible. The HIPed product's microstructure is finer than the forged product's. Fig. 8 shows the microstructure photographs of the surface area and the internal area of the HIP product.

Hot Isostatic Pressing – HIP'17                                      Materials Research Forum LLC
Materials Research Proceedings **10** (2019) 58-64          doi: http://dx.doi.org/10.21741/9781644900031-8

There were no differences between the surface area (after capsule removal) and the internal areas.

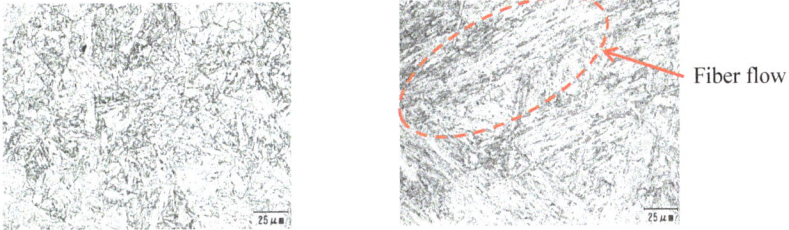

Fiber flow

*Fig. 7 The microstructure of the HIPed product (left) and the forged product*

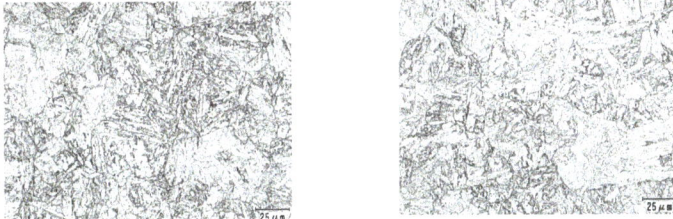

*Fig. 8 The microstructure of the surface area (left) and the internal area (right) of the HIP product*

| Table 4 Comparison of a HIP-NNS product and a forged product | | | |
|---|---|---|---|
| | Forged | HIP-NNS | Reduction ratio (%) |
| Material weight [ kg ] | 900 | 536 | 40 |
| Machining time [ hours ] | 100 | 70 | 30 |

The non-metallic inclusion measurement was conducted according to ASTM E 45 A. In the HIPed product there is a small amount of grained oxide but there is no sulfide, alumina, or silicate and as such it is very clean for a HIPed product. These results show that there is no difference in mechanical properties between the outside and the inside of the HIPed product.

**Machining time and material weight comparison**

A comparison of the machining time for the forged product and the HIP-NNS product was conducted using the following fomula. Reduction of machining time = (|machining time from forged block - machining time from HIP NNS product |) / machining time from forged block x 100%

The fomula for a material weight comparison between the forged product and the HIP-NNS product was calculated by the following formula. Reduction of material weight = (|forged block weight - HIP NNS product weight |) / forged block weight x 100%

Table 4 below shows the comparison result. The forged material weight is 900 kilograms in contrast to the much lighter HIPed NNS at 536 kilograms. The machining time for a forged product is 100 hours but the HIPed NNS product is as low as 70 hours. This means that the

HIPed NNS product can offer us a 40% reduction in material weight and a 30% reduction of machining time compared to a forged product.

## Conclusions

In conclusion we can see that the difference from the overall target size was *only* 3.2 to 5.5 percent and this was mainly due to the capsule design and construction. We can also see that we achieved a better than ASTM standard for all related tests. Thirdly the HIPed NNS product's microstructure is homogeneous across all thicknesses which means there is almost no difference in the mechanical properties throughout the product. And significantly the manufacturing of a valve body using HIP NNS technology provides a 40% reduction in material weight and a 30% reduction of machining time compared to a forged product underlining the benefits to both the environment and the manufacturer.

## References

[1] David W. Gandy, John Siefert, Lou Lherbier, Dave Novotnak "PM-HIP RESEARCH, APPLICATIONS, AND TECHNOLOGY GAPS FOR THE ELECTRIC POWDER INDUSTRY" Proceedings of International Conference on HIP, 2014, pp. 130-150.

[2] T. Shiokawa, Y. Yamamoto, S. Hirayama and Y. Nagamachi "Comparison of experimental and FEM simulations of densification during HIP processing of powder into a cylindrical component" Proceedings of International Conference on HIP, 2011, pp. 225-230.

Hot Isostatic Pressing – HIP'17
Materials Research Proceedings **10** (2019) 65-72

Materials Research Forum LLC
doi: http://dx.doi.org/10.21741/9781644900031-9

# Experience in HIP Diffusion Welding of Dissimilar Metals and Alloys

V.N. Butrim [1,a,*], A.G. Beresnev [1,a], V.N. Denisov [1,a], A.S. Klyatskin [1,a],
D.A. Medvedev [1,a], and D.N. Makhina [1,a]

[1] Joint-Stock Company "Kompozit", 4 Pionerskaya Str., Korolev 141070,
Moscow Region, Russia

[a] info@kompozit-mv.ru

**Keywords:** Welding, Hot Isostatic Pressing, Diffusion Bonding, Dissimilar Metals

**Abstract.** HIP solid-state diffusion welding is a controlled production operation at all the processing stages. Unlike other known solid-state welding techniques the HIP allows to provide strong and dense bonding with stability properties irrespective of the sizes and a configuration of the contact surfaces of materials welded. Here we present some special pilot examples of HIP diffusion welding of dissimilar metals and alloys: steel XM19-to-steel 316L, bronze Cu-Cr-Zr–to-steel 316L, copper M1–to-steel Fe-18Cr-10Ni-Ti-C, titanium alloy Ti-6Al-4V–to-steel Fe-18Cr-10Ni-Ti-C, single-crystal molybdenum-to-polycrystal molybdenum and titanium alloy-to-aluminum alloy.

## Introduction

Solid-state diffusion welding (DW) is a main way to make a bimetallic structural material for space and nuclear application where a strong and dense bonding of materials with different chemical compositions is needed. This technology produces a monolithic joint resulting from a maximum closing of the contact surfaces due to their local plastic deformation at the increased temperature as well as the formation of metallic bond at the atomic level followed by a mutual diffusion of the components through the surface layers of the materials bonded [1]. Solid-state diffusion welding includes the following obligatory stages: the oxide film removal from contact surfaces, the actual contact formation, the surfaces activation, the chemical bond formation and diffusion. This sequence is true for all known methods of solid-state welding: cold bonding, explosion welding, percussion vacuum welding, friction welding, vacuum roll welding, induction and ultrasonic welding, etc. However, only the diffusion welding is the most universal and reliable method that allows controlling all four key technological parameters of process: temperature, pressure, dwell time and diffusion medium. The method of diffusion welding (DW) with use of hot isostatic pressing (HIP) can be considered as a kind of classical DW wherein technological parameters can be controlled within a wider range. Below we examined the influence of the HIP DW technological parameters on a welded joint quality.

## Influence of HIP parameters

*Temperature and pressure*

Temperature and pressure are mutually dependent parameters in HIP technology. Specified pressure values in a HIP installation chamber are achieved by thermal expansion of working gas as the temperature increases. Thus, with computation of the necessary amount of gas at the room temperature performed, it is possible to reach the HIP operation conditions both in the temperature of 200 °C to 1200 °C and pressure of 20 MPa to 200 MPa ranges under any parameter combination. As the pressure is created by gas, the pressure value will be the same in any point of the HIP product contact surfaces despite the sizes and configurations of the parts bonded. As it is well known [2], if all-round compression pressure is applied to a crystal the concentration of vacancies in this case will then be equal to

$$Cp = Co \; exp \; (-P\Omega/kT),$$
(1)

Where, $Co$ is the equilibrium component concentration at $P=0$; $\Omega$ is the atomic volume; $P$ is the all-round compression pressure; $k$ is the Boltzmann's constant; $T$ is the temperature. In this case the "minus" symbol denotes compression. That is, the amount of vacancies decreases with increase in pressure, such that the diffusive flow of atoms decreases too. In 1954, S. Storchheim et al. [3] established that the phase $Ni_3Al_2$ was not formed even at pressure of 170-300 MPa, only the phase $Ni_3Al$ was formed at pressure higher than 300 MPa, and intermetallic phases were not observed at a pressure about 500 MPa. Thus it is possible to increase or reduce diffusion rate with pressure increasing or decreasing. In so doing it is possible to reach such process conditions wherein the nucleation and growth of undesirable phases can be depressed at the contact zone.

*Dwell time*
Theoretically the duration of a HIP DW technological parameters can be unlimited and depends only on the end result required. HIP DW excludes the void volume along a boundary of the dissimilar metal diffusion bonding that caused by distinction in partial component diffusion coefficients, for example, nickel and copper (Kirkendall's effect [3]), as owing to constantly applied pressure the formed vacancies are replaced with metal atoms having the largest diffusion velocity, here copper (Fig.1). Therefore, it is possible to create quite a wide transitional area in a contact zone of dissimilar metals (up to several hundred microns) by operating of HIP DW duration. Increasing the transitional area width will give the positive effect, for example, as damping layers between metals of greatly different coefficients of thermal expansion.

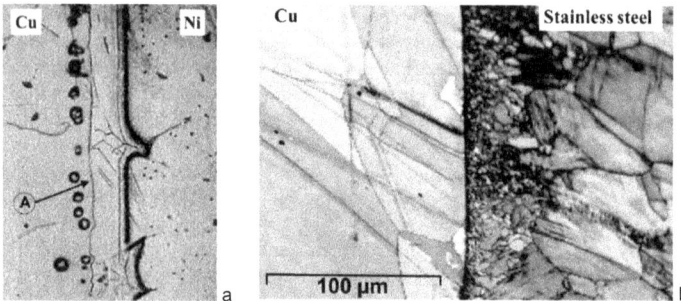

*Fig.1 – Voids in copper of Ni-Cu diffusion bonding according to Le Claire A.D. and Barnes R.S. [3], A is the initial line of contact (a), absence of voids after HIP DW [4] (b)*

*Environment*
Under the fine vacuum and at the high temperatures the dissolution of oxides promotes the formation of juvenile contact surfaces of the joints welded.

## Experimental procedure

The following materials are used in this study: steel XM19 (chemical composition in wt%: 22 Cr, 12 Ni, 5 Mn, 2.5 Mo, 2.5 Nb, 0.2 V and the reminder Fe) in forging, steel 316L (in wt%: 17 Cr, 12 Ni, 2.5 Mo and the reminder Fe) in forging, stainless steel in wt%: 18 Cr, 11 Ni, 0.5 Ti and the reminder Fe in bar and sheet, bronze in wt%: 0.9 Cr, 0.1 Zr and the reminder Cu in sheet, titanium alloys in wt%: 6 Al, 4 V and the reminder Ti in bar; 4.5 Al, 5 V, 2 Mo, 1.2 Cr, 0.6 Fe and the reminder Ti in sheet, copper alloy M1 in bar, aluminum alloy in wt%: 6 Mg, 0.7 Mn and the reminder Al in sheet, single-crystal and especially pure polycrystal molybdenum in bars. To manufacturing of samples for test of mechanical properties and research of structure used one HIP diffusion bonding from party, and in a design of structural assembly of the diverter and mirrors were provided with special places for cutting of samples witnesses. Mechanical tensile strength testing was carried out according to requirements of the ISO 6892:1984, ISO 783:1989, ISO 783-89 standards. Microstructure was observed by of an optical microscope Zeiss Axio Observer with ImageExpert system and a raster electronic microscope JSM-6610LV equipped with Advanced AZtec EDS Detector. Metallographic samples were made with use of a combination of the machines which includes the Delta AbrasiMet Abrasive Cutter, SimpliMet 3000 Mounting Press and EcoMet 250 Grinder-Polisher.

## Results

*Steel XM19-to-steel 316L HIP Diffusion Bonding*

Within an International Thermonuclear Experimental Reactor (ITER) program the diffusion welding has been performed of large-size parts of corrosion-resistant stainless steel AISI 316L and high-strength steel XM19 intended for pre-fabrication of the diverter attachment fitting (Fig. 2a). The structural assembly mass is equal to 760 kg and the summary diffusion bonded surface area is nearly 770 cm$^2$ (Fig. 2b) and 1260 cm$^2$ (Fig. 2c). Failure of the bimetallic tension specimens takes place on the base metal of steel 316L outside the diffusion bonding zone (Fig. 3a) as tensile strength of the HIP diffusion bonding zone is higher than tensile strength of steel 316L. In microstructure of the diffusion bonding zone steel XM19-to-steel 316L (Fig.3b) the presence of oxide phases is not detected. Besides, the common grains were observed in a contact area.

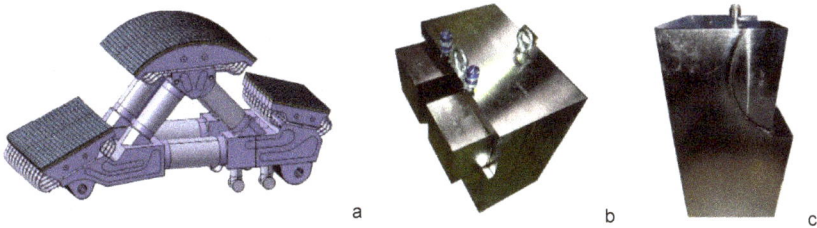

*Fig.2 - Drawing of the diverter attachment fitting (a) and photographs of the bimetallic steel XM19-to- steel 316L HIP DW assemblies (b, c).*

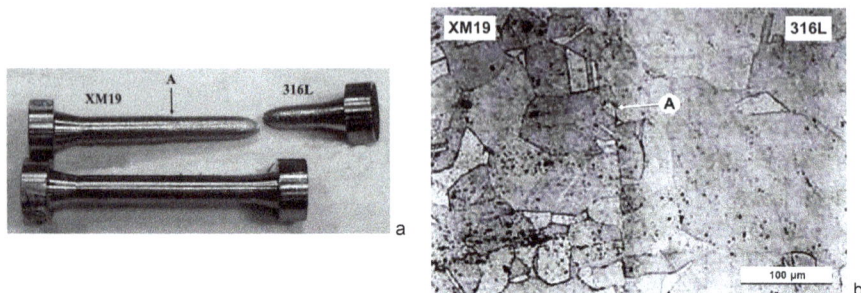

*Fig. 3 – The bimetallic tension specimens steel XM19-to-steel 316L before and after tensile testing (a), micrograph of a transitional layer (b), A – contact area [4].*

*Bronze Cu-Cr-Zr-to-steel 316L HIP Diffusion Bonding*
A bimetallic bronze Cu-Cr-Zr-to-steel 316L heat exchanger of the first ITER wall is a complicated design with internal chambers and cooling channels (Fig. 4). Various welding processes for making pressure-tight contact surfaces of assembly parts prior to HIP are tested: manual argon-arc fusion welding; electron beam welding and automatic laser welding; the vacuum brazing and generally accepted one with the use of the capsule. Microstructure of the transitional layer of the HIP DW bronze-to-steel joint has much the same character despite the area and curvature of the contact surface. Thickness of a visible transitional layer is about 7 μm. A chemical composition of this layer contains elements characteristic of both for steel and bronze, and the chrome content here is higher than in the steel. The diffusion depth of copper into the steel reaches 30 μm beyond the transitional layer boundary. Diffusion depths of iron, nickel and chromium from steel into bronze are 50-100 μm, 40-60 μm and 5-7 μm, correspondingly. In bronze, at a distance of 1-3 μm from the transitional layer boundary a chain of inclusions takes place of up to 1μm and increased zirconium content, the nature of its formation being not determined (Fig.5).

*Fig.4 – The models of bimetallic bronze Cu-Cr-Zr-to-steel 316L heat exchanger of the first ITER wall with internal chambers (a) and cooling channels (b) [4]*

The post-HIP model of the first ITER wall was subjected to heat treatment in bronze standard mode: water hardening from temperature 980°C with the subsequent aging. After heat treatment tensile failure of bimetallic samples witnesses takes place on the main component of an alloy of bronze and average values of strength made 420 MPa at room temperature and 350 MPa at temperature of 250 $^0$C.

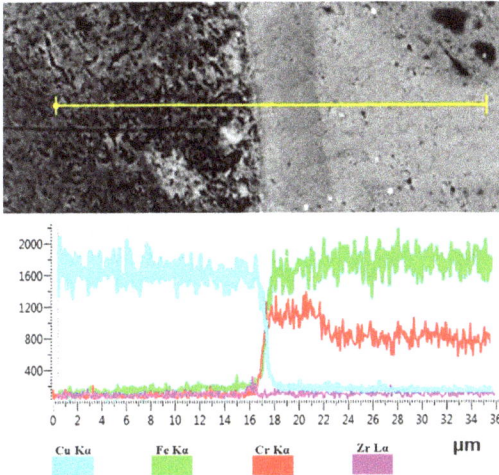

*Fig.5 – Typical electron micrograph of bronze/SS HIP DW joint, pink line shows the region of alloying element spectrum*

*Copper / stainless steel HIP Diffusion Bonding*

Copper/stainless steel adapters (Fig.6a) were made of 100 mm-diameter HIP DW-bimetallic bar-piece blanks followed by machining. The capsules each have 4 pairs of the piece blanks. The analysis of the microstructure testifies that the transitional layer of the HIP DW copper-to-stainless steel joint has 100% density (Fig.6b). Bending test of a 7x30x80-mm test piece cut from the bimetallic blank did not lead to its failure (Fig.6c)

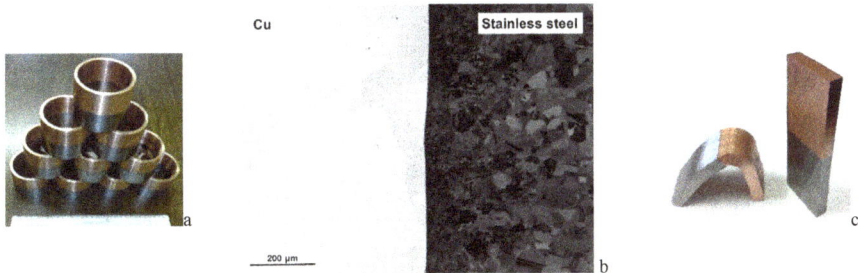

*Fig. 6 – Copper-to-stainless steel adapter (a); micrograph of the transitional layer (b); a 7x30x80 mm test piece before and after bending tests (c) [4].*

*Ti6Al4V alloy / stainless steel HIP Diffusion Bonding*

Direct Ti6Al4V alloy / stainless steel HIP DW does not give positive results. With increase in iron concentration more than 0.1 wt%, intermetallic TiFe and $TiFe_2$ are formed in titanium alloys, which embrittles the DW transitional layer. In practice [5], niobium and copper spacer-

Hot Isostatic Pressing – HIP'17                                     Materials Research Forum LLC
Materials Research Proceedings 10 (2019) 65-72          doi: http://dx.doi.org/10.21741/9781644900031-9

foils are recommended for use in this case. Niobium and titanium form a continuous number of the solid solutions, and therefore, DW of them is not difficult. However, when reacted with carbon from steel niobium forms carbides along the total contact area, which embrittles the diffusion bonding transitional layer too. Carbon-tight, the copper spacer-foil located between niobium and steel does not allow forming niobium carbides. The relation between copper and niobium foil thicknesses is of 1.5 − 3.0 thereby initiating the failure on the base metal of the copper spacer-foil, which is responsible for stable strength properties and good ductility of the bimetallic joint obtained [6].

Ti6Al4V alloy / Fe18Cr10Ni1Ti stainless steel adapters (Figure 6a) were made of 50 mm-diameter HIP bimetallic bar-piece blanks followed by machining. The capsules each have 6 pairs of the piece blanks (Fig.7a). All the HIP diffusion bonding contact zones have 100%-densities and no ply separations and pores are observed. The diffusion width of the titanium-to-niobium contact zone is up to 35 μm (Fig.7b). The diffusion bonding width of the niobium to copper contact zone is up to 6μm without visible structural changes.

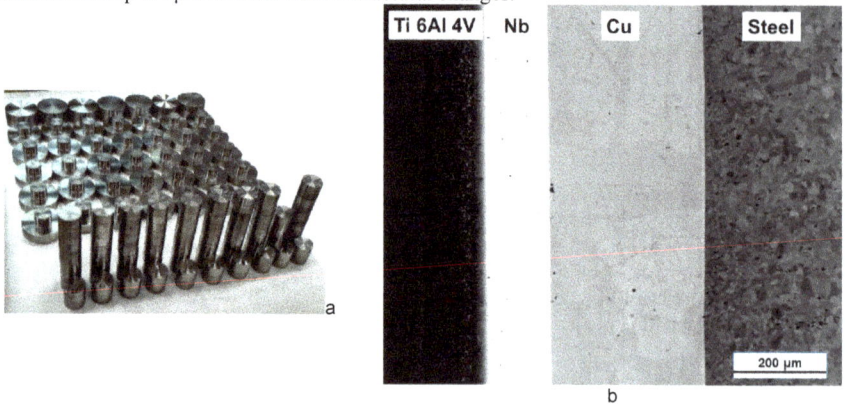

*Fig.7 - Titanium-steel piece blanks (a), micrograph of the Ti6Al4V/Nb/Cu/SS joint [4] (b)*

With diffusion from steel into copper, iron and nickel atoms diffuse into copper at the depth of up to 15 μm and 5 μm, respectively, of the visible contact line. The depth of copper diffusion into steel is observed up to 5 μm only. Thus the common diffusion width of copper-to-steel DW transitional layer reaches 20 μm. Room temperature tensile strength of the HIP diffusion bonding Ti6Al4V alloy-to-stainless steel is equal to 439 MPa. Ti6Al4V / stainless steel adapters have passed hydrostatic test at pressure of 7000 MPa / 700 bar. No leakage occurred.

*Single-crystal molybdenum / polycrystalline molybdenum HIP Diffusion Bonding*
Single-crystal molybdenum ($Mo_{single}$) is material of choice for producing the first mirrors of 200 mm x100 mm to be used in a diagnostics system for ITER Hydrogen Lines Spectroscopy [7]. Both the severe quality requirements and the complicated 250 mm-diameter blank technology determine very high price of the molybdenum single crystals. Therefore, a composite mirror design has been proposed. In this design, the reflecting part, produced from several parts of single-crystal molybdenum with the same orientation in the crystallographic planes, is joined by

Hot Isostatic Pressing – HIP'17                                    Materials Research Forum LLC
Materials Research Proceedings **10** (2019) 65-72         doi: http://dx.doi.org/10.21741/9781644900031-9

HIP DW with the extra pure polycrystalline molybdenum ($Mo_{poly}$) of a relatively low price. The technology of joining the molybdenum single crystal with the base made of polycrystalline molybdenum should not cause recrystallization of the single-crystal molybdenum. The mechanical properties of the diffusion welded joint should be higher than the loads applied while mirror manufacturing (milling and turning, grinding, and polishing, etc.) and servicing. The HIP diffusion bonding between single-crystal molybdenum and polycrystalline molybdenum was achieved by the titanium foil interlayer of 0.1 mm thickness ($Ti_{0.1}$) [7]. The titanium and molybdenum form a continuous number of the solid solutions and therefore there is no danger of forming any embrittlement phases in the contact zone. Room temperature tensile strength of the HIP diffusion bonding $Mo_{single}/Ti_{0.1}/Mo_{poly}$ shows more than 380 MPa. Failure happens on a titanium foil interlayer and has viscous character. Loss of the mirror heat conductivity due to a titanium interlayer is less than 5 %. A sharply defined contact line and a visible homogeneous light transitional zone of ~ 5-10 µm thickness being a solid solution of molybdenum in titanium is observed on the both sides of the titanium foil interlayer within a diffusive zone of the $Mo_{single}/Ti_{0.1}/Mo_{poly}$ bonding (Fig. 8a). The composite single-crystal molybdenum mirrors with working surface area of 5000 $mm^2$ and 8000 $mm^2$ (Fig.8b) have successfully passed the tests carried out according to a special program, including tests in the conditions of heating to a temperature of 250 °C under hydraulic pressure up to 50 bar.

Fig.8 - *Micrograph of* $Mo_{single}/Ti_{0.1}/Mo_{poly}$ *bonding (a) and experimental composite single-crystal molybdenum mirrors(b) [7]*

*Titanium alloy / aluminum alloy HIP Diffusion Bonding*
The fusion welding of bimetallic titanium alloy/aluminum alloy is impossible because intermetallic $TiAl_3$ and $TiAl$ are formed in alloying zone at 1340 °C and 1460 °C, respectively. Ultimate solubility of titanium in aluminum is as low as 0.26-0.28 wt% and 0.07 wt% at 665 °C and at the room temperature, respectively. HIP DW allows for obtaining the titanium alloy-to-aluminum alloy bonding at a temperature of 500-560 °C without formation of the intermetallic in the contact zone. To increase the thin-walled bimetallic design serviceability it has been suggested that on the surface of titanium part a relief can be carried out as a thread profile having an identical radius of curvature equal to ½ height of the thread ledge with a base size equal to doubled height of the thread ledge [8]. The role of the relief is to obtaining a more developed surface in the contact zone and its activation due to intensity of the local shear deformation along the profile thread ledges and hollows and formation of the physical contact of the metals during HIP, which finally leads to an increase in mechanical properties and tightness. The existence of curvatures causes lack of stress concentration. The positive effect is reached when the relief

Hot Isostatic Pressing – HIP'17                                                                    Materials Research Forum LLC
Materials Research Proceedings **10** (2019) 65-72                           doi: http://dx.doi.org/10.21741/9781644900031-9

period quantity is not less than two of them on the bimetallic design wall thickness (Fig.9a). Shear strength of the samples with relief is 119MPa, which is twice higher than that of the joint without relief ($\tau$s =58 MPa). Intermetallic phases are not detected in the contact zone microstructure (Fig.9b).

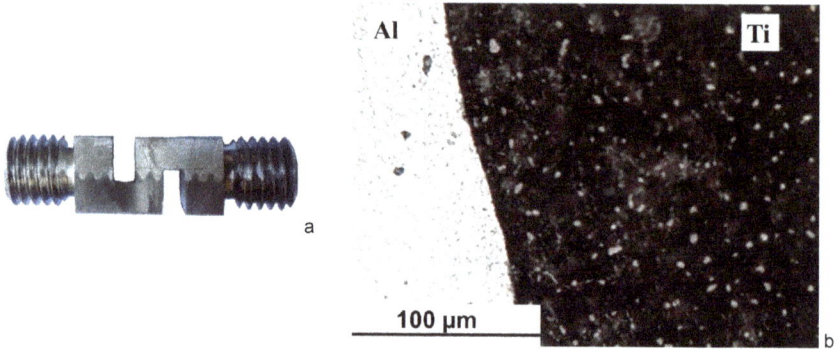

*Fig.9 – Bimetallic Ti-Al shear-test specimen (a), a micrograph of cross section of Ti-Al joint (b)*

## Conclusion

The HIP solid-state diffusion welding is a controlled production operation at all the processing stages. Unlike other known solid-state welding techniques the HIP allows for providing the strong and dense welded joint of stability properties despite the area and configurations of the contact surfaces of materials welded.

## References

[1] G. B. Konyushkov, R. A. Musin, Special Methods of Pressure Welding, IPR-Media, Saratov,2009, in Russian.

[2] B.S. Bokstein, Diffusion in Metals, Metallurgy, Moscow, 1978, in Russian.

[3] W. Seith, Diffusion in Metals, Foreign Literature, Moscow, 1958, in Russian.

[4] A.S. Klyatskin, V.N. Denisov, V.N. Butrim [etc.], HIP Diffusion Welding, Perspective materials, 11, (2011) 362-369, in Russian.

[5] N. F. Kazakov, Diffusion Welding of Metals, Mashinostroenie, Moscow, 1981, in Russian.

[6] V.N. Denisov, A.S. Klyatskin, V.N.Butrim [etc], RU Patent 2612331 (2017).

[7]. V.N. Denisov, A.S. Klyatskin, V.N.Butrim [etc,] Producing Composite Single-Crystal Molybdenum Mirrors by Diffusion Welding in the Hot Isostatic Pressing Conditions, Welding International, 7, (2017), 1-6

[8] V.N. Denisov, A.S. Klyatskin, V.N.Butrim [etc], RU Patent 2620402 (2017).

Hot Isostatic Pressing – HIP'17
Materials Research Proceedings 10 (2019) 73-78

Materials Research Forum LLC
doi: http://dx.doi.org/10.21741/9781644900031-10

# Heat Treatment inside the HIP Unit

Nils Wulbieter[1,a *], Anna Rottstegge[1,b], Daniel Jäckel[1,c], Werner Theisen[1,d]

[1]Chair of Materials Technology, Ruhr-Universität Bochum, Germany

[a]wulbieter@wtech.rub.de, [b]anna.rottstegge@reiloy.com, [c]jaeckel@fgw.de,
[d]theisen@wtech.rub.de

**Keywords:** HIP, Heat Treatment, Phase Transformation, Hardenability, Latent Heat, Electrical Conductivity

**Abstract.** The possibility of combining densification or compaction of steel parts with a heat treatment has recently evolved due to the production of HIP units with a rapid quenching device. Several studies have already been performed to assess the cooling speed and show possibilities for heat-treating steels. It has already been shown that several alloyed steel grades could be hardened by quenching inside a HIP unit. This study aims to characterize the impact of high isostatic pressure during austenitization and quenching on the transformation behavior and resulting microstructure of hardenable steels. The effect of pressure during quenching was studied using two methods. The first method is to measure the latent heat inside the transforming steel during isothermal holding. The release or uptake of energy reveals information about the transformation sequence taking place. The second method is to use the electrical resistivity of a steel as a sensitive indicator for the existing phases and solution state of the steel during continuous cooling after austenitization. Both experimental methods reveal that an isostatic pressure of 170 MPa is sufficient to shift the transformations to longer times and lower temperatures and hence increase the hardenability of hardenable martensitic steel.

## Introduction

A HIP unit was recently introduced that offers the opportunity to quench inside the pressure vessel [1]. Since the introduction of this URQ[TM] method (Uniform Rapid Quenching), various research teams have investigated the possibility to heat-treat steel inside a HIP unit. Mashl showed that hardening of a low-alloyed steel inside a pressure vessel leads to higher hardness compared to quenching in oil [2]. The same result was found by Weddeling [3,4]. Angré et al. have shown that a pressure of 170 MPa prolongs pearlite formation in steel specimens that are austenitized and subsequently hold isothermally in the pearlite region [5].

These findings show that a high isostatic pressure of 170 MPa does have an influence on the phase transformations of steel. Therefore, the TTT (Time-Temperature-Transformation) diagrams of every steel are not applicable for HIP heat treatment. In order to correctly predict the hardness and microstructure resulting from an integrated heat treatment in the HIP unit, the TTT diagram as well as the pressure effect must be known. At ambient pressure, the measurement of variations in length during phase transitions (dilatometric measurements) is utilized to determine the temperature and time of a phase transition; however, this was not possible inside a HIP+URQ[TM] unit.

In the present study, two methods that are capable of indicating phase transitions in theory are tested in a HIP unit and evaluated for two steels. One method, which also was utilized by Angré et al., is to measure the latent heat. The emission or absorption of heat is an indication for a phase transition. The second method of determining a phase transition is to measure the electrical conductivity. It is known from the literature that changes in the electrical conductivity during heating or cooling can be related to phase transformations or carbide precipitation [6,7]. Thus, it is of interest to evaluate the method of measuring the electrical conductivity as an indicator for a phase transition during quenching of steels inside a HIP unit.

## Experimental

*Materials*
The latent heat was measured using a cylinder made of AISI H13 (DIN X40CrMoV5-1), and the electrical conductivity was measured using AISI W210 (DIN 100V1). The chemical composition in mass-% of the materials is given in Table 1.

*Table 1: Chemical composition of the investigated steels.*

| [mass-%] | C | Si | Mn | Cr | Mo | V |
|---|---|---|---|---|---|---|
| AISI H13 | 0.37 | 0.93 | 0.32 | 4.88 | 1.28 | 0.92 |
| AISI W210 | 1.0 | - | 0.22 | - | - | 0.1 |

*Heat treatment in the HIP unit*
The heat treatment was performed inside a hot isostatic press QIH9 with URQ$^{TM}$ from Quintus Technologies AB. Technical details are given in [4]. The highest pressure at which quenching is possible is 170 MPa. The effect of pressure was analyzed at 25 and 170 MPa; 25 MPa was chosen as a comparatively low pressure that still offers a reasonable quenching efficiency. The heating rate was chosen to be 40 K/min, and heating and pressurizing took place simultaneously. Cooling rates could be changed in three steps (fast, medium, slow) by reducing the volumetric flow rate of the gas by changing the gas inlet nozzle.

*Measurement of the latent heat during isothermal holding*
The latent heat was measured using a cylinder of H13 (Ø 70 x 20 mm) with a drilled hole for a core thermocouple (see Fig.1).

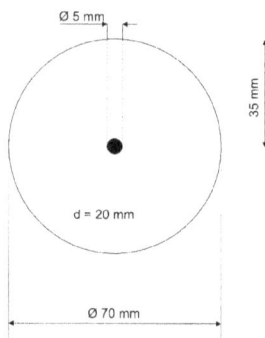

*Fig. 1: Geometry of the H13 cylinder for measuring the latent heat. The black circle shows the position of the core thermocouple.*

Two further thermocouples were used to control the furnace temperature. The material was austenitized at 1050 °C, held for 30 minutes, and quenched at maximum speed to the isothermal holding temperature. The holding temperatures ranged from 710 °C to 790 °C in increments of

10 K. According to an isothermal TTT diagram of H13 at ambient pressure, the pearlite transformation takes place in this temperature region. All trials were run directly one after the other without changing the specimen or moving the thermocouples.

*Measurement of the electrical conductivity*
Another method of determining an in-situ phase transformation is to measure the electrical conductivity as a function of the cooling temperature. A phase transformation leads to a significant change in the measured electrical conductivity. The temperature and corresponding time at which the transformation takes place are shown by the intersection point of two tangents (Fig. 2).

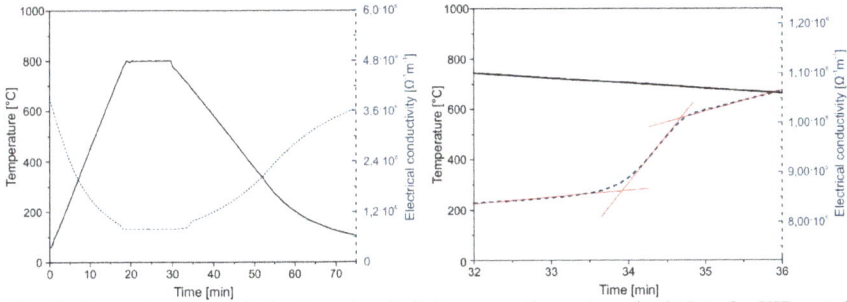

*Fig. 2: Determination of the begin and end of phase transformation of W210 in the HIP unit due to changes in the electrical conductivity.*

Steel W210 was used for this investigation because of its comparatively early pearlite transformation. Preliminary tests showed that the materials used for this measurement must have a significant length to increase the signal intensity. Therefore, wound W210 wire was used. The wire was 300 mm in length and 1 mm in diameter. The feedthroughs for thermocouples were reused as feedthroughs for measuring the electrical conductivity. To exclude the contact and conductor resistances, we opted for the four-wire technique. The electrical conductivity was measured and recorded using the nanovolt-/micro-ohmmeter Keysight 34420 A in combination with BenchVue Digital Multimeter Pro software. The contact points for introducing the current were the ends of the specimen wire, the contact points for measuring the voltage were 10 mm apart from the ends. Contacts were made with spot welding. Further technical details are given in [4]. Quenching trials inside the HIP were performed with two different pressures (25 MPa, 170 MPa) and three different cooling rates. The begin and end of this phase transformation, measured by changes in the electrical conductivity, were compared to a continuous TTT diagram of W210 at atmospheric pressure.

**Results and Discussion**
*Pressure-induced delay of isothermal pearlite transformation of H13 measured via the latent heat*
Fig. 3a shows the core temperature of the H13 cylinder during quenching and isothermal holding at 760 °C under two different pressures (25 MPa, 170 MPa). It can be seen that the pearlite transformation takes place due to a significant increase in temperature during holding at 760 °C,

which results from the latent heat. Furthermore, it can be seen that the maximum of the temperature peak decreases with increasing pressure and is shifted to longer times. The peak temperature correlates to the heat released during the phase transformation of austenite to pearlite. It can be concluded that the amount of released heat correlates to the amount of formed pearlite. This result shows the slowing down and delay of pearlite transformation under pressure. Whereas the begin of the phase transformation can not be measured precisely due to strong undercooling at high pressures, the end of the transformation can be precisely measured as a function of the holding temperature. In Fig. 3b, the end time of transformation is plotted against the holding temperature, which ranges from 700 °C to 770 °C. It can be seen that the end of phase transformation is clearly shifted to longer times.

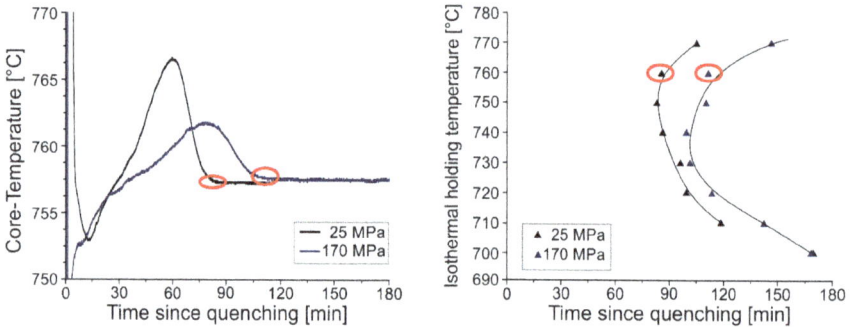

*Fig. 3: a) Measured core temperature of the test cylinder of H13 as a function of pressure at a constant furnace temperature of 760 °C. The red marks indicate the end of phase transformation. b) Shifted phase transformation under pressure at different holding temperatures.*

This effect can be explained by the influence of pressure on the thermodynamic equilibrium according to further investigations at high pressure [8,9]. Therefore, a high pressure stabilizes the phase with a lower molar volume and higher density, respectively. In this case, austenite has a lower molar volume compared to martensite. This leads to an expansion of the austenite phase field and shifts the transformation line to lower temperatures. Therefore, stronger undercooling is needed to initiate pearlite transformation, while at the same time, the diffusion of elements during pearlite transformation slows down due to the lower temperature. The results can be verified with the corresponding microstructure. Fig. 4 shows the resulting microstructure after quenching and holding at 770 °C for 1 hour for 25 MPa (A) and 170 MPa (B). Sample B shows significantly smaller amount of pearlite compared to sample A, which is in agreement with the results of the latent heat measurements (Fig. 3). Furthermore, the microstructure of sample B is clearly finer. Thus, improved mechanical properties from HIP heat-treated samples are expected and will be the object of further investigations.

*Pressure-induced delay in continuous pearlite transformation of W210 measured by electrical conductivity*
Fig. 5a shows the cooling curves of W210 wire measured for three different cooling speeds at 25 MPa plotted into a TTT diagram of W210 at atmospheric pressure. The marked points represent

the begin and the end of pearlite transformation measured by changes in the electrical conductivity. The measured transformation points are in sufficient agreement with the pearlite transformation curves in the TTT diagram. In contrast, Fig. 5b shows that under a high pressure of 170 MPa and high cooling rates, the phase transformation is shifted to lower temperature compared to the conventional TTT diagram. The reason for this is the same as that discussed in the previous section. The high pressure stabilizes the austenitic phase because of its lower molar volume and higher density, respectively [8,9]. Currently, from a technical point of view, it is not possible to vary the cooling speed in the QIH9 HIP unit in more steps than shown in the diagram. However, with further cooling steps it would be possible to create a TTT diagram under pressure for the steel of interest. For this reason, the measurement of electrical conductivity is a promising method to measure the begin and end of a phase transition in a HIP unit under pressure.

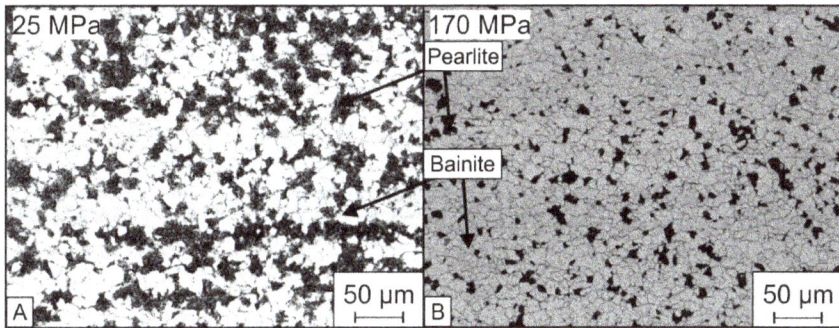

*Fig. 4: Resulting microstructure of H13 after quenching from 1050 °C to 760 °C, holding for 1 hour and further quenching to room temperature under a pressure of (A) 25 MPa and (B) 170 MPa.*

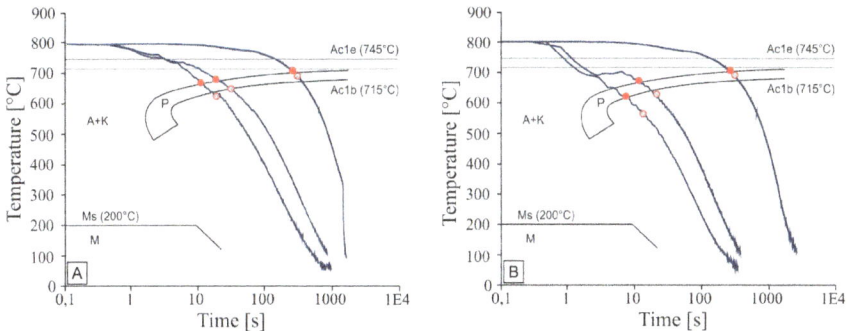

*Fig. 5: Cooling curves of W210 as a function of three different cooling speeds and a pressure of (A) 25 MPa and (B) 170 MPa, plotted in a TTT diagram valid for ambient pressure. The marked points indicate the measured phase transition due to changes in electrical conductivity.*

## Conclusions and Summary

Measuring the core temperature of massive parts and the electrical conductivity of wire specimens offers the possibility of detecting phase transitions during heat treatment inside a hot isostatic press. It was shown that both methods can be used to monitor the austenite-to-pearlite transformation as a function of pressure in the HIP unit. Additionally, it was shown that a HIP pressure of 170 MPa is suitable for shifting the pearlite transformation to longer times and lower temperatures. Measurement of the latent heat is a comparatively simple method of determining isothermal TTT diagrams under HIP pressure. However, for the H13, the begin of pearlite transformation could not be measured precisely due to strong undercooling during quenching. Enhanced adjustment control could solve this issue. In contrast, it is possible to determine the begin and end of pearlite transformation by measuring the electrical conductivity. This leads to the possibility of generating TTT diagrams under pressure in HIP units.

The results of this investigation also show that HIP units that can quench with very high cooling rates are an interesting tool for the heat treatment of materials. This HIP heat treatment leads to better hardenability and a finer microstructure of the examined steels.

Further research needs to be undertaken to investigate the mechanical properties after such a HIP heat treatment.

## References

[1] Ahlfors, M.: "The Possibilities and Advantages with Heat Treatments in HIP", Proceedings of the 11th Conference on hot isostatic pressing, Stockholm (2014).

[2] Mashl, S.J., Eklund, A. and Ahlfors, M.: "Evaluation of Ultra-High Gas Pressure Quenching During HIP", Proceedings of the 28th ASM Heat Treating Society, Detroit (2015).

[3] Weddeling A., Wulbieter, N., Theisen, W.: "Densifying and hardening of martensitic steel powders in HIP units providing high cooling rates. Powder Metallurgy 59 (1) (2016) pp. 9-19. https://doi.org/10.1080/00325899.2015.1109803

[4] Rottstegge, A.: "Strukturbildungsprozesse von Eisenbasislegierungen beim heißisostatischen Pressen", PhD thesis, Ruhr-Universität Bochum (2017).

[5] Angré, A., et al.: Phase transformation under isostatic pressure in HIP. Powder Metallurgy 60 (3) (2017) pp. 167-174. https://doi.org/10.1080/00325899.2017.1318479

[6] Watson, T.W., Flynn, D.R. and Robinson, H.E.: "Thermal Conductivity and Electrical Resistivity of Armco Iron", Journal of Research of the National Bureau of Standards 71C (4) (1967) pp. 285-291.

[7] Klein, S. et al.: "Diffusion process during cementite precipitation and their impact on electrical and thermal conductivity of a heat-treatable steel", J Mater Sci 52 (2017) pp. 375-390. https://doi.org/10.1007/s10853-016-0338-1

[8] Nilan, T.G.: "Morphology and kinetics of austenite decomposition at high pressure", Transactions of the Metallurgical Society of AIME 239 (6) (1967) pp. 898-909.

[9] Kakeshita, T., Saburi, T., Shimizu, K.: "Effects of hydrostatic pressure and magnetic field on martensitic transformations", Materials Science and Engineering A 273-275 (1999), pp. 21-39. https://doi.org/10.1016/S0921-5093(99)00287-7

Materials Research Forum LLC
doi: http://dx.doi.org/10.21741/9781644900031-11

# HIP Activities for Turbopump Components of Korea Space Launch Vehicle

Suk-Hwan Yoon[1,a]*, Chang-Ho Choi[1,b] and Jinhan Kim[1,c]

[1]Korea Aerospace Research Institute, 169-84 Gwahak-ro, Yuseong-gu, Daejeon 34133, Korea

[a]shyoon@kari.re.kr, [b]cch@kari.re.kr, [c]jhkim@kari.re.kr

**Keywords:** Aerospace, Launch Vehicle, Liquid Rocket Engine, Turbopump, Impeller, Turbine, Blisk

**Abstract.** In Korea, we are developing liquid rockets for commercial launch services, and the government agency, Korea Aerospace Research Institute (KARI), is responsible for main development. Turbopump, which is a key component of liquid rocket engine, is a rotating machine that pressurizes fuel and liquid oxygen in an extreme environment and supplies them to a combustion chamber. Design requirements are very severe because it must maintain lightweight feature while outputting very large power. The HIP (Hot Isostatic Press) method is a near-net shape processing, which makes it easy to mold a material that is difficult to machine, while securing quality comparable to forged products. These advantages are particularly attractive for the aerospace sector. Recently, we tried manufacture of turbopump impellers and turbine discs using HIP technology, and some of the products have been assembled in a turbopump and ground-tested. This will be described in detail in this paper.

## Introduction

There is a series of space launch vehicle programs in Korea and they are named KSLV (Korea Space Launch Vehicle) programs [1]. The first program, KSLV-I was successfully launched in Jan. 2013 after two launch failures. The first stage of KSLV-I was developed in Russia and the upper stage was covered in Korea. Now KSLV-II program is in progress and the launch is scheduled in 2020. The vehicle is composed of three stages, the first stage with four 75 ton thrust engines, the second stage with a single 75 ton thrust engine, and the third one with a 7 ton thrust engine [2]. All of the vehicles are under development by Korea Aerospace Research Institute (KARI) in Korea, requiring precedent development of 75 ton thrust and 7 ton thrust liquid engines. The both engines employ pump-fed gas generator cycle with kerosene/LOx, and the turbopump consists of single-stage centrifugal pumps for each propellant and a single-stage impulse turbine in one axis. An Inter-Propellant-Separator (IPS) is installed between the oxidizer pump and the kerosene pump to avoid any interaction between propellants [3]. Fig. 1 shows 75 ton and 7 ton turbopumps under development in KARI. They completed performance tests and were assembled to engines, now the engines are undergoing ground performance tests.

Various materials are utilized to fabricate turbopumps. Especially heat resistant nickel alloys are widely used in oxidizer pump and turbines due to their excellent mechanical properties at extremely low or high temperature condition. These superalloys usually have poor machinability so that casting and powder sintering methods are known to be suitable fabrication methods. In the turbopumps of KSLV-II, the impeller of the oxidizer pump is of Inconel 718 alloy, and is manufactured by machining and brazing process. Also the turbine blisk is of the same material and is manufactured by electric discharge machining and turning operation. In this paper manufacturing of these items with hot isostatic pressing was investigated. For the turbine blisk, near-net shape processing with hot isostatic pressing was tried on the billets. And for the impeller, consumable cores with low carbon steel were used to make fluid passages through leaching process.

*Fig. 1: 75 ton and 7 ton turbopumps developed by KARI*

**Manufacture of HIP billets for turbine blisk**
The first step of introducing HIP process in the KSLV turbopump was manufacture of HIP billets for turbine blisk. Before that the blisk was made by cutting/turning and EDM machining from a forged Inconel 718 billet, and the mechanical machining processes were time consuming steps. Therefore near-net shape process with HIP operation was tried to minimize mechanical machining steps, leaving only EDM and final machining process.

*Fig. 2: Canned metal powder before HIP(left), after HIP(center) and machined disk(right)*

*Fig. 3: Blades and shroud made by EDM machining*

Hot Isostatic Pressing – HIP'17                                    Materials Research Forum LLC
Materials Research Proceedings **10** (2019) 79-84          doi: http://dx.doi.org/10.21741/9781644900031-11

Fig. 2 shows HIP process to make a billet for turbine blisk. After a blisk is obtained by HIP, EDM machining is applied from the both sides to implement blades and shroud as shown in Fig. 3. In the beginning, manufacturing billets of plain cylindrical disk was the main focus so that substantial amount of post process machining was required before the EDM process. Nowadays initial can shapes are being optimized to minimize subsequent machining.

**Manufacture of Impellers**

Impellers for oxidizer pump are also made from Inconel 718 due to its excellent cryogenic properties. Conventionally machining/brazing or casting methods are applied to these parts. Fig. 4 shows two impellers made with machining/brazing and investment casting.

*Fig. 4: Machined/brazed impeller (left) and investment casted impeller (right)*

It is well known that consumable, sacrificial metal core has to be used for HIP manufacture of shrouded impellers [4, 5]. At first step, fabricating simplified sample was tried with low carbon steel and nitric acid. The core and the sample are shown in Fig. 5.

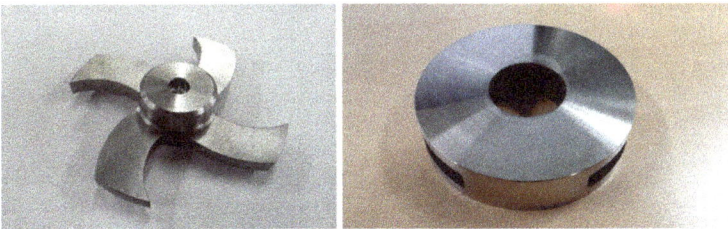

Fig. 5: Simplified, low carbon steel core and sample specimen

After sample trial, impeller of the oxidizer pump was selected as the next target. The fluid passage between blades has complex three-dimensional shape and that core making should start from the 3D modelling of it, as represented in Fig. 6. After the modelling process, a five-axis machining center was used to machine the core. As for the cans, they are of axisymmetric shape and can be machined by plain tuning operation. Plain low carbon steel was used for the core material, and STS 316 alloy for the cans.

Hot Isostatic Pressing – HIP'17                                    Materials Research Forum LLC
Materials Research Proceedings **10** (2019) 79-84          doi: http://dx.doi.org/10.21741/9781644900031-11

Machined core and cans before HIP process are shown in Fig. 7.

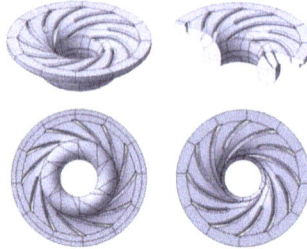

*Fig. 6: 3D model of the consumable core*

*Fig. 7: Machined core and cans*

After core and cans are assembled, Inconel 718 powder was filled up, and then the assembly was welded and evacuated before entering into the HIP furnace. HIP was done under temperature of 1160°C and pressure of 100 MPa for 4 hours. After HIP process the low carbon steel core was leached using nitric acid. The rest processes are final machining and coating to complete the impeller, as shown in Fig. 8.

*Fig. 8: Core leached and machined impeller*

Hot Isostatic Pressing – HIP'17                                    Materials Research Forum LLC
Materials Research Proceedings **10** (2019) 79-84          doi: http://dx.doi.org/10.21741/9781644900031-11

It is also known that carbon is diffused from the core to the impeller surface due to highly different carbon contents between low carbon steel and nickel base heat resistant alloys. Therefore it is recommended to remove a few microns of diffusion layer with some removal processes, such as chemical milling.

*Fig. 9: Cut and leached impeller for inspection*

Manufactured impellers were cut and inspected as shown in Fig. 10. The impeller was cut between hub and shroud, before leaching process. After leaching, blade and hub profiles were measured using a 3D coordinate measuring machine. Several cycles of trial and error for core design were necessary until satisfactory results were obtained.

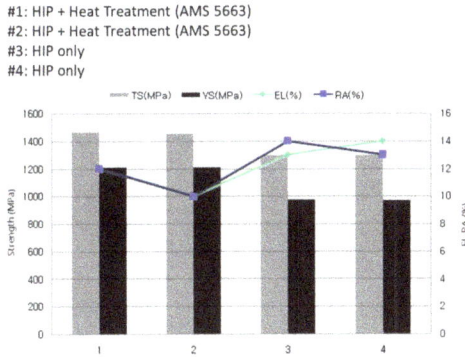

*Fig. 10: Mechanical properties of HIP Inconel 718 specimens*

Mechanical properties of HIP Inconel 718 specimens were measured using a tensile testing machine. The results are shown in Fig. 10, which complied with the AMS 5663 specification.

**Conclusion**

Manufacture of the billets for turbopump turbine blisk was performed using Inconel 718 powder. In the beginning a bulk cylindrical disk was manufactured, but can shapes are being optimized to minimize subsequent machining process. With the billets obtained, EDM and final machining were applied to complete the manufacture of the turbine blisk.

Besides the turbine blisk, impellers for oxidizer pump were tried using leaching process of consumable core. Low carbon steel core was modeled, machined and then inserted into cans for HIP process. HIP was performed with Inconel 718 powder and with the core, which was later leached by nitric acid to implement blades and shroud of the impeller. The impellers were cut through the blades for inspection and measure, several cycles of core modeling and HIP were necessary to achieve satisfactory dimensional accuracy. Also HIP Inconel 718 specimens were prepared to measure the mechanical properties. They showed good strength and toughness results, in compliance with the AMS 5663 specification.

## References

[1] J. Kim, Status of the development of turbopumps in Korea", Journal of the Korean Society of Propulsion Engineers, Vol. 12, No. 5 (2008) 73-78.

[2] J. Ko, S.Y. Cho, Space Launch Vehicle Development in Korea Aerospace Research Institute, Proceedings of the 14th International Conference on Space Operations (2016) 2530. https://doi.org/10.2514/6.2016-2530

[3] J. Kim, E. S. Lee, C. H. Choi and S. M. Jeon, Current status of turbopump development in Korea Aerospace Research Institute, Proceedings of the 55th International Astronautical Congress, Vancouver (2004) IAC-04-S.P.17

[4] V. Samarov, C. Barre and D. Seliverstov, Net Shape HIP for complex shape PM parts as a cost efficient industrial technology, Proceedings of the 8th International Conference on Hot Isostatic Pressing, Paris (2005) 48-52.

[5] C. Bamton, W. Goodin, T. V. Daam, G. Creeger and S. James, Net-Shape, HIP powder metallurgy components for rocket engines, 8th International Conference on Hot Isostatic Pressing, Paris (2005) 1-10.

Hot Isostatic Pressing – HIP'17                                 Materials Research Forum LLC
Materials Research Proceedings **10** (2019) 85-91      doi: http://dx.doi.org/10.21741/9781644900031-12

# HIP Processing of Improved Tooling Materials for High-Productivity Hot Metal Forming Processes

Maxime Gauthier[1,a *], Guillaume D'Amours[2,b] and Fabrice Bernier[1,c]

[1] Automotive and Surface Transportation Research Centre, National Research Council Canada
75, de Mortagne Blvd., Boucherville (Québec), Canada, J4B 6Y4

[2] Aluminum Technology Center, National Research Council Canada 501, Université-Est Blvd.,
Saguenay (Québec), Canada, G7H 8C3

[a] Maxime.Gauthier@cnrc-nrc.gc.ca, [b] Guillaume.Damours@cnrc-nrc.gc.ca,
[c] Fabrice.Bernier@cnrc-nrc.gc.ca

**Keywords:** Hot Isostatic Pressing, HIP, Tool Steels, Metal Matrix Composites, Hot Working, Hot Stamping, Hot Metal Forming, Automotive, Aunch, Tooling, Thermal Conductivity

**Abstract.** Much work has been carried out in the last decade on the development of high performance alloys to reduce vehicle weight. These alloys are often characterized by low room-temperature formability. A variety of hot forming processes (hot stamping, hot extrusion and high-pressure die casting) are thus being used or adapted for these alloys. The final mechanical properties, shape complexity and production cost of parts made using these processes will be closely related to mold/die thermal and mechanical performance.

Hot work tool steels generally have the required mechanical properties and durability to meet hot-processing requirements but have low thermal conductivity. The stringent low processing cost and high-volume production requirements of the automotive industry compel part producers to find ways to shorten unit production times at equivalent product quality. In order to meet the processing requirements of advanced alloys and transfer heat more rapidly, the tooling should thus have a higher thermal conductivity than the standard tool steel dies currently in use.

The aim of this work is to optimize die properties to improve heat transfer kinetics during part shaping, thus providing an increase in efficiency and productivity for automotive metal part manufacturing. Hot Isostatic Pressing (HIP) has been used to clad a conformal-cooled copper core with a layer of either hot-work tool steel or High-Thermal Conductivity (HTC) composite material designed at NRC. Properties and performance of these systems are compared with those of standard tool materials to demonstrate the practical potential for future development and optimization of advanced tooling.

## Introduction

A promising way to manufacture structural automotive components using high strength AA7xxx aluminum alloy sheets is through the hot stamping process. The process itself is not new and is currently used in production with boron steel sheets. Rana *et al.* [1] provides different process details with boron steel and the average processing time is about 7 min for each part. For AA7xxx aluminum alloy sheets, the hot stamping processing sequence can be summarized as follows: 1) a blank sheet is heat treated to put the alloying elements into solid solution, 2) the blank is rapidly transferred to the press, 3) the punch is partially closed to shape the part and 4) the punch is completely closed to quench the shaped part and prepare the alloy for precipitation 5) the part is removed from the press and artificially aged to reach high mechanical strength. The hot stamping process with aluminum is similar to steel except that the solution heat treatment step is longer and aluminum's thermal conductivity is higher. During the last few years, aluminum alloys for hot stamping have attracted the interest of many scientists. Quenching rate effects for AA7xxx alloys have been analyzed by Keci *et al.* [2] and Kumar *et al.* [3]. High temperature mechanical behavior and high temperature formability analysis and modeling have been analyzed by Mohamed [4] and Elfakir [5]. Harrison *et al.* [6] have also produced real AA7xxx aluminum alloy pillars using hot stamping.

Due to the major investments required for future part production, another vital process parameter to consider is the hot stamping cycle time. During blank quenching, heat is transferred to the punch, the latter being cooled either by flowing water or oil via internal cooling channels. Conventional tool steels are used for the punches, yet their thermal conductivity is low, which increases the cycle time. As other authors have also realized [7, 8], the development of HTC tool steels would thus contribute to the improvement of hot working efficiency and productivity, which would be beneficial for the automotive industry where large production volumes require low processing times and costs.

**Experimental**

HIP processing and characterization of reference tool steel and HTC composite : Spherical powders of D2 tool steel powder (-150+45 μm, Sandvik – see composition in Table 1) and of pure copper (grade 153A, 2%max+100μm / bal+45μm / 10%max-45μm, ACuPowder) were used to process the materials required for this study. D2 is not generally used for hot working, but it was chosen as it has been a reference for different sheet forming studies at NRC during the last few years, so comparisons with earlier work could easily be made (smaller-scale preliminary work based on H13 tool steel gave similar results [9]). For the production of the D2 reference and the development of the HTC tool steel, cylindrical 304L stainless steel canisters (190 mm-high, 138mm OD, 1.59 mm wall thickness) were filled with either D2 powder or a D2+30vol.% Cu blend and tapped to tap density. A cover plate featuring a tube for gas evacuation was welded on top of each of the canisters, which were then submitted to a vacuum degassing treatment (14h @ 150°C, 4h @ 550°C under mechanical vacuum (~7x10$^{-2}$ Torr)). After mechanical crimping of the vacuum tubes and sealing by TIG welding, each canister was then HIPed in a model AIP10-30H hot isostatic press from American Isostatic Presses, Inc. The HIP parameters chosen for the pure D2 material were the following: 4h @ 1100°C and 15000 PSI (103 MPa). In the case of the D2+30vol.% Cu blend, the HIP plateau temperature was decreased to 1000°C to avoid formation of a liquid Cu phase. No additional heat treatment was applied to the resulting HIPed materials.

*Table 1: Chemical Composition (wt. %) of D2 Powder*

| Fe | Cr | C | Mo | V | Mn | Si | Ni | P | S | Cu |
|----|----|----|----|----|----|----|----|----|----|----|
| Bal. | 12.8 | 1.41 | 0.95 | 0.72 | 0.6 | 0.27 | 0.19 | 0.02 | 0.01 | 0.01 |

After HIPing, coupons were machined out of the D2 and D2+30vol. % Cu billets by wire Electro-Discharge Machining (w-EDM). These specimens were used for evaluation of Heat Capacity ($C_p$) by Differential Scanning Calorimetry (NETZSCH DSC 404F3), Thermal Diffusivity ($\alpha$) by Laser Flash Analysis (NETZSCH LFA 457 Microflash) and Hardness (Instron Series B2000). Thermal Conductivity (k) was calculated using the measured $C_p$ and $\alpha$ values by means of the following relationship, where $\rho$ is the density of the material:

$$\alpha = \frac{k}{\rho c_p} \tag{1}$$

HIP processing of hot stamping punches: The processing of conformal-cooled, hemispherical punches followed similar processing steps and HIP parameters as were used for the tool steel and HTC composite processing. Three punches were processed: 1) D2 reference, 2) 2mm-thick D2 layer HIP-clad on a solid Cu core and 3) 2mm-thick D2+30vol.% Cu layer HIP-clad on a solid Cu core - see Fig. 1a) to c). HIP temperatures were 1100°C for the D2 reference and 1000°C for the Cu-containing punches.

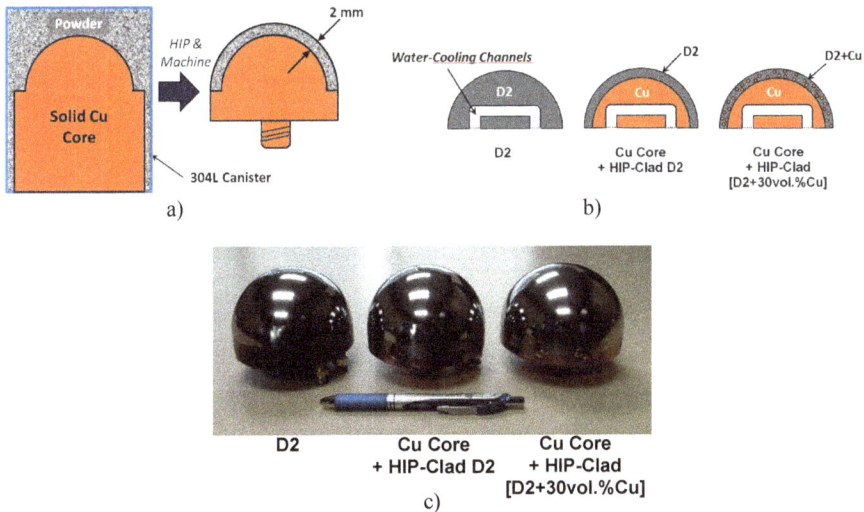

*Figure 1:*    *Conformal-Cooled, Hemispherical Punches:*
       *a)*   *Schematic cross-section of canister with core and powder used to prepare HIP-clad punches;*
       *b)*   *Schematic cross-sections of the dome part of punches processed by HIP;*
       *c)*   *Hemispherical punches processed by HIP*

Laboratory-scale aluminum hot stamping process: Study of hot stamping of AA7xxx sheets using these punches was carried out using a laboratory scale, hot-stamping test setup designed and installed on a 100 kN MTS hydraulic system (Fig. 2). With this particular test equipment, it was possible to reproduce the same steps during the same period of time as in production, except that some manual operations had to be carried out. During each test, a hot blank (sheet) was placed in the binder that was then partially closed; the punch then rapidly deformed the blank to a specific position. Heat was partially transferred to the binder but mainly to the punch due to higher contact pressures. The punch was water-cooled and three, Omega 20-mil exposed

Hot Isostatic Pressing – HIP'17                                    Materials Research Forum LLC
Materials Research Proceedings **10** (2019) 85-91        doi: http://dx.doi.org/10.21741/9781644900031-12

thermocouples with thermal conductive paste were introduced by the bottom side up to very near the upper punch surface.

The D2 tool steel punch was first used for ten identical hot stamping tests. The cycle time was constant and adjusted to let the punch cool down to a steady-state temperature. Afterwards, the punch with HIP-clad D2 over Cu was used and ten additional tests were repeated for the same cycle time. Finally, the punch with HIP-clad D2+30%Cu over Cu was tested using the same procedure.

*Figure 2:      Four-inch Nakazima set-up for hot stamping trials.*

**Results**

Characterization of D2 tool steel reference and HTC composite: The calculated thermal conductivity values at relevant temperatures are shown in Figure 3a. It is seen that the thermal conductivity of the HTC composite is significantly higher than that of the D2 base steel at all temperatures (close to double the D2 reference value). This proves that the strategy of adding Cu to D2 to improve k gave good results. In the case of hardness (Fig.3b), as was to be expected, the hardness of the Cu-containing material is much lower than that of the pure D2 counterpart.

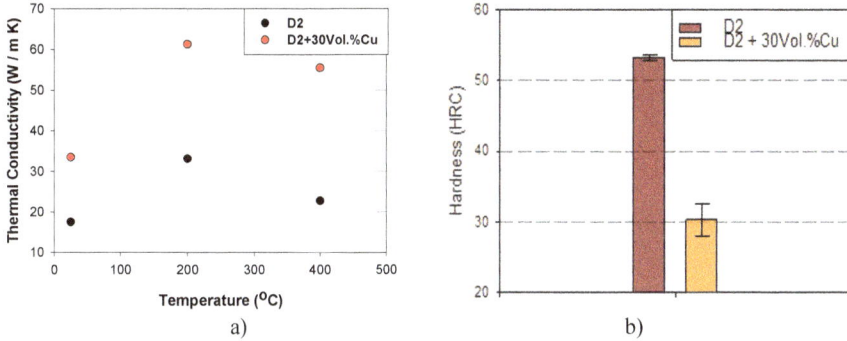

Figure 3:   a)   Thermal conductivity of D2 and HTC composite at 25, 200 and 400 °C;
            b)   Room temperature hardness of D2 and HTC composite;

Nakazima hot stamping results: Hot stamping trials of AA7xxx sheets were repeated ten times with each punch. The first tested punch was the one made of D2 reference tool steel. Fig. 4 shows the punch temperature for the last (tenth) stamping test. The acquisition time was six minutes per test and an interval of one minute between each test was maintained. The water temperature in the cooling lines was less than 283 K (10°C).

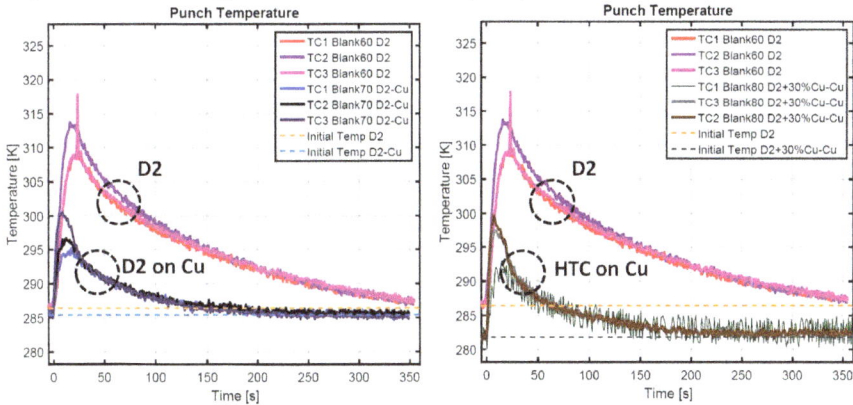

Figure 4:   Thermocouple temperatures for the three different punches

Fig. 4 shows that the D2 steel punch temperature (upper three curves – both graphs are identical for D2 values) did not reach thermal equilibrium, even with more than six min of rest. The maximum temperature measured by the thermocouples for the D2 steel punch for this particular test was 314 K (41°C), which occurred 16 sec after the punch made contact with the blank. This shows the time required for heat to be transferred into the D2 tool steel, even if punch thermocouples were close to the surface in contact with the aluminum alloy sheet. The last 300 sec of acquisition were used to extrapolate the D2 temperature curves up to a line corresponding to the initial punch temperature, with the use of an exponential function. This procedure was also repeated for the four tests preceding the tenth one in order to generate average values for a total of five tests.

*Table 2: Time required to reach punch temperature equilibrium after hot stamping*

| Die material | Mean time to reach equilibrium (s) | Fraction of D2 reference mean time to reach equilibrium (%) | Gain in Efficiency (%) |
|---|---|---|---|
| D2 | 381 | - | - |
| D2 over Cu | 181 | 48 | 110 |
| D2+30%Cu over Cu | 168 | 44 | 127 |

The slopes of the exponential extrapolations at the intersection with the line representing the initial punch temperatures were also recorded and scaled by the differences between the maximum punch temperatures, as recorded by the thermocouples, and the initial punch temperatures. These slopes can be used to represent the heat transfer and the cooling rate at that particular moment of a given punch temperature vs. time curve. When these slopes become small, this suggests that thermal equilibrium is close to being reached. For the two other punch materials, the same procedure was applied to find the time needed to reach thermal equilibrium. The punch temperature curves for these new materials are also shown in Fig. 4: "D2 on Cu" (left-hand graph) and "HTC on Cu" (right-hand graph). A significant difference in heat transfer behavior is observed for these two materials when compared to that of the D2 steel. Essentially, they are showing lower maximum values, showing the higher heat transfer rates of the "HIP-clad/copper core" punches. The slopes of these curves were also analyzed, normalized and compared to the normalized slopes obtained for the D2 steel. The cooling time at which the normalized slopes were equal was used to determine the start of thermal equilibrium; these mean time values are presented in Table 2 for each punch material. It can be seen that the HIP-clad D2+30%Cu on a copper core leads to the lowest cooling time (only 44% of the cooling time of the D2 reference punch). The gain in efficiency values of the two improved punches are seen to be >100% and were computed using:

$$Gain\ in\ Efficiency = \left\{\left[\frac{(Cycle\ time\ D2)}{(Cycle\ time\ improved\ die)}\right] - 1\right\} \times 100 \qquad (2)$$

**Conclusion**

A HTC composite, hot-work tooling material was successfully processed by HIP. Its thermal conductivity was improved over the D2 reference, while its hardness was lower.

Based on these available thermal and mechanical properties, development work on an optimal HTC composite should continue.

Conformal-cooled, hemispherical punches were manufactured by HIP and used for hot stamping of AA7xxx sheets. Die cooling performance was as follows: HTC on Cu > D2 on Cu > D2. Gains in efficiency were of 127% for HIP-clad HTC on Cu and 110% for HIP-clad D2 on Cu. A thin, HIP-clad layer of HTC tool steel on a conformal-cooled copper core significantly contributes to productivity increase and should be considered for automotive sheet forming.

## References

[1]  R. Rana, S.B. Singh, Automotive Steels, Design, Metallurgy, Processing and Applications, Woodhead Publishing, first ed., (2017).

[2]  A. Keci, N. Harrison, S. Luckey, Experimental Evaluation of the Quench Rate of AA7075, SAE Technical Paper 2014-01-0984, (2014).

[3]  M. Kumar, N.G. Ross, Investigations on the Hot Stamping of AW-7921-T4 Alloy Sheet, Advances in Materials Science and Engineering (2017) 567-573. https://doi.org/10.1155/2017/7679219

[4] M.S.K. Mohamed, An Investigation of Hot Forming Quench Process for AA6082 Aluminium Alloys, PhD thesis, Imperial College London, 2010.

[5] O. Elfakir, Studies on the Solution Heat Treatment, Forming and In-Die Quenching Process in the Production of Lightweight Alloy Components, PhD thesis, Imperial College London, 2015.

[6]  N. R. Harrison, S.G. Luckey, Hot Stamping of a B-Pillar Outer From High Strength Aluminum Sheet AA7075, Sae International Journal of Materials & Manufacturing 7(3) (2014) 567-573. https://doi.org/10.4271/2014-01-0981

[7] Klein, S.; Weber, S.; Theisen, W., European Conference on Heat Treatment 2016 and 3rd International Conference on Heat Treatment and Surface Engineering in Automotive Applications (2016)

[8] Wilzer, J.; Küpferle, J.; Weber, S.; Theisen, W. steel research int. (steel research international) (2015), Volume 86, Issue 11, Pages 1234–1241

[9] F. Bernier, M. Gauthier, unpublished NRC work.

# Recent Developments of HIP Equipment in JAPAN

Megumi Kono[1,a]* Katsumi Watanabe[1,b]

[1]Kobe steel Ltd., Takasago, Japan

[a] kono.megumi@kobelco.com, [b] watanabe.katsumi@kobelco.com

**Keywords:** HIP, Rapid Cooling, High Productivity

**Abstract.** Toll services have been increased recently in the Japanese HIP market. This trend leads to larger HIP equipment and shorter cycle times for productivity improvement. In addition, longer life cycle of pressure vessels are demanded to reduce the costs in conformance with the requirements of the relevant laws and regulations of Japan.

To meet such demands, KOBELCO has developed a new rapid cooling system and the first product was delivered in 2016. This new cooling system ensures a rapid cooling rate while achieving the design life cycle by low design temperature of the pressure vessel. At the development stage of the new cooling system, the numerical analysis of the heat flow during rapid cooling was conducted using new techniques including a real gas model and a new model for the thermal insulator. This article will introduce this new rapid cooling system and describe other related topics.

**Introduction**

Since KOBELCO started the sales of HIP equipment in the 1970s, we have developed many types of equipment for many applications. For example, for cemented carbide parts in 1971, high speed tool steel billets in 1977, and soft ferrite in the 1980s, and so on. There were two very important development challenges. One was "Diversification of the process atmosphere", and the other was "High productivity."

*Table 1. Typical development accomplished by KOBELCO [1-2]*

| | |
|---|---|
| Small R&D high pressure gas equipment | 1964 |
| Production HIP for cemented carbide parts | 1971 |
| Production HIP for high speed tool steel billets | 1977 |
| Production HIP with bottom loading system | 1977 |
| Production HIP for soft ferrite | 1978 |
| Nitrogen HIP unit | 1981 |
| Modular HIP system for soft ferrite production | 1982 |
| Oxygen HIP unit for R&D of ceramics | 1986 |
| Oxygen HIP unit for commercial production | 2002 |
| Oxygen partial pressure control HIP | 2002 |
| Mechanical properties testing equipment in hydrogen | 2003 |
| New Rapid cooling method | 2015 |

**Development challenges**

The first development challenge was "Diversification of the process atmosphere."

In the beginning, the basic gas atmosphere for HIP has always been argon. In Japan there were many requirements for various gases for various processes.

Hot Isostatic Pressing – HIP'17                                    Materials Research Forum LLC
Materials Research Proceedings **10** (2019) 92-97        doi: http://dx.doi.org/10.21741/9781644900031-13

What is in production today is:

Nitrogen gas for silicon nitride used for ceramic ball bearings

Oxygen + Argon for superconductive electrical wire

Shroud HIP to prevent damage from gas generated in the product

KOBELCO has always tried to find better way to handle various gases. And these HIPs are still in use today and play an important part in their respective industries.

Another development challenge was "High productivity."

HIP equipment must have a thick-walled pressure vessel for high pressure. This makes HIP equipment more costly than commonly used heat treatment equipment. Moreover, raising and lowering the temperature takes time, and in very large HIP equipment, one HIP cycle may take a whole day or more to complete. In order to address such issues, productivity improvement was required to reduce the processing cost per workpiece. From the very beginning, KOBELCO has designed the bottom loading system for high productivity. In the 1980s the Modular HIP system was incorporated into HIP equipment. In 2015 a newly developed rapid cooling furnace was installed to a considerable degree of high productivity.

**Modular HIP system**

In the 1980s, a modular HIP system was incorporated into HIP equipment in order to achieve full density in the production of high quality soft ferrite. A modular HIP system has two or three cooling stations and preheating stations as shown in Fig. 1.

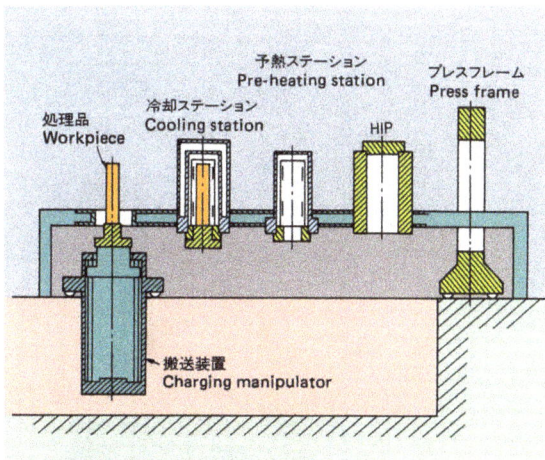

*Fig. 1. Modular HIP system*

The workpiece can be transferred while it is kept in an inert atmosphere furnace. It allows the preheated workpiece to be inserted into the pressure vessel for HIP treatment, or taken out in a hot state after HIP treatment and transferred to the cooling station to cool down. As a result, only the pressurizing, holding, and depressurizing processes are performed in the HIP pressure vessel, leading to 2 to 2.5 times higher productivity.

**KOBELCO Old Rapid cooling system**
In the global market, rapid cooling furnaces that directly cool the high temperature and high pressure hot zone were made commercially available, significantly contributing to the reduction of processing time. KOBELCO also worked on the development of a rapid cooling furnace. In the 1990s, we produced a prototype and conducted a performance verification test. Unfortunately, however, we could not put it into commercial production. At that time, ASEA had a patent on wire-wound, interior cooling vessels which provided the most efficient cooling performance, and competitors including KOBELCO had to use liner type cooling vessels for small and middle size units and mono block vessels incorporating a rapid cooling furnace for large size units. Mono block vessels with a low interior cooling performance allow an excessive temperature rise at the pressure vessel inner surface due to direct rapid cooling gas. To prevent such excessive temperature rise, a heat sink was required inside the pressure vessel to dissipate heat from the hot zone. A mechanism for keeping the hot zone at a uniform temperature in rapid cooling was also needed. Therefore, to install these two systems into the pressure vessel, the very low volumetric efficiency of the pressure vessel impeded cost effective treatment and the full utilization of rapid cooling properties. Consequently, ASEA had a monopoly in the direct, rapid cooling HIP market.

**New rapid cooling furnace**
To solve the problem, KOBELCO has developed a new rapid cooling furnace. Fig.2 shows the diagrammatic illustration of the new rapid cooling furnace.

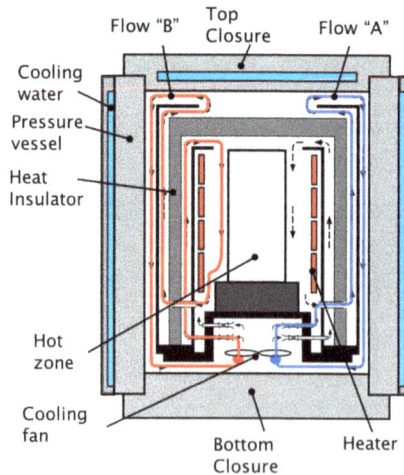

*Fig. 2. New rapid cooling furnace*

The heat insulator which surrounds the hot zone, blocks heat transfer and provide thermal control when the temperature in the hot zone gets high. The new rapid cooling furnace has vertical cylinders inside and outside of the heat insulator to direct the medium gas upward. These cylinders feature circulation flows which are generated inside and outside of the heat insulator.

Materials Research Forum LLC
doi: http://dx.doi.org/10.21741/9781644900031-13

The gas blown by the cooling fan circulates, flowing between the outer cylinder and the heat insulator up to the top of the heat insulator, then between the heat insulator and the pressure vessel down to the cooling fan while being cooled along the pressure vessel inner wall. This is called "first circulation flow," shown as "Flow A" in Fig.2.

Then the gas blown by the cooling fan flows between the inner cylinder and the heat insulator up to the top of the hot zone: From there, it flows into the hot zone to cool the hot gas, and then merges with the first circulation flow to circulate back to the cooling fan. This is called "second circulation flow," shown as "Flow B" in Fig.2.

Rapid cooling by using these two circulation flows is the one of the biggest features of the new furnace.

As you may know if you are familiar with fluid mechanics, the higher the gas flow rate, the higher the heat exchange rate between the gas and metal on the metal surface. The commonly used rapid cooling furnace roughly uses the second circulation flow only to perform rapid cooling. In this case, the cooling rate in the hot zone depends on the amount of the cooling gas in the hot zone.

The hot gas passing through the hot zone is directed to the pressure vessel inner surface, thus it is necessary to reduce the gas flow rate to a certain level or lower in order to prevent an excessive temperature rise at the pressure vessel inner surface. That means the maximum gas flow rate is limited. On the other hand, in the new rapid cooling furnace, the first circulation flow does not pass through the hot zone, enabling an increase of the gas flow rate regardless of the cooling rate. The second circulation flow is cooled while merging with the first circulation flow that prevents an excessive temperature rise at the pressure vessel inner surface and allows an increase of the first circulation flow rate. As a result, a heat exchange is performed between the relatively cool mass flow of gas and the pressure vessel inner surface.

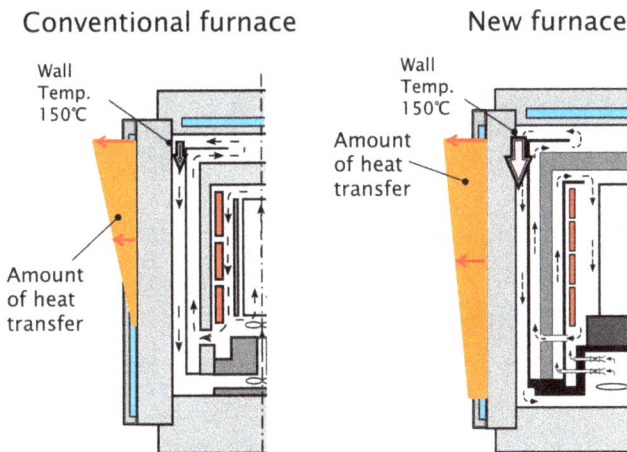

Fig. 3. Comparison of two different furnaces

The sufficiently increased heat exchange rate between the gas and the pressure vessel inner surface, allows effective use of the entire pressure vessel inner surface for heat exchange to prevent an excessive temperature rise at the pressure vessel inner surface, and makes best use of the heat extraction capability of the pressure vessel. Another feature of this new rapid cooling furnace is that the hot zone does not require any additional devices such as a circulation fan since the temperature in the hot zone can be kept constant. This is made possible by introducing the second circulation flow from the top of the hot zone.

Fig.3. shows a comparison of two different furnaces. The design of the wire-wound interior cooling vessel is the same in both cases. The left drawing shows a typical conventional rapid cooling furnace with only the second circulation flow. The right drawing shows the new rapid cooling furnace with the first and second circulation flows. The orange area shows the calculated amount of heat transfer. In either case, the gas flow rate is set so that the inner wall temperature of the vessel is uniform (150 °C). The gas is argon and the temperature of the hot zone is 1100 °C. As Fig.3 shows, the rapid cooling with the first and second circulation flows on the right side can absorb more heat than the rapid cooling with only the second circulation flow on the left side. It is necessary to discharge more heat from the hot zone through the pressure vessel to rapidly lower the temperature in the hot zone. This new rapid cooling furnace can discharge more heat from the pressure chamber to achieve faster cooling of the hot zone, compared to a conventional cooling furnace.

As a result of calculating the cooling rate of the hot zone inside HIP according to previous drawings, the new rapid cooling furnace is 1.5 times faster than conventional furnaces. In this case, a comparison was made with the same temperature on the inner surface of the pressure vessel. However, if the cooling rate is fixed, the new rapid cooling furnace can achieve rapid cooling with a relatively low temperature of the pressure vessel.

The life of the pressure vessel depends on the design conditions such as the thickness of the vessel, the design temperature, the conditions of the wire winding, etc. In Japan, in particular, since it is necessary to use the maximum value of the inner surface temperature as the design temperature of the pressure vessel, designing a lower temperature can ensure a longer vessel life. The same can be said about design pressure. Therefore, with the new rapid cooling furnace, the vessel life can be designed to be longer, or the design pressure can be higher.

### The first commercial unit of new rapid cooling furnace

In 2016, the first commercial HIP equipment employing this new rapid cooling furnace was delivered to the customer. Table 2. shows the main specifications of this HIP equipment – among the largest scale in our product line. We have developed wire-wound, interior cooling vessels in parallel to the development of the rapid cooling furnace. This large HIP equipment has our largest interior cooling vessel with piano wire wound around the cylindrical core using a spacer. A heater has been designed and developed by incorporating the new rapid cooling furnace to meet the specifications shown in Table.2. and installed in this HIP equipment.

*Table 2. Main specifications of the first commercial unit.*

| Hot zone diameter | 850mm |
|---|---|
| Hot zone height | 2,500mm |
| Maximum temperature | 1400 °C |
| Maximum pressure | 147MPa |
| Maximum weight of work load | 4,500kg |
| Cooling rate with no load | 40 to 60°C/min |
| Cooling rate 2ton load | 15 to 25°C/min |

Fig.4. shows the results of the rapid cooling performed at the cooling rate of 15°C/min with approximately 2-ton load using this HIP equipment. As shown in this graph, the rapid cooling was performed smoothly in the range of 1,150°C down to about 400°C at the cooling rate of an approximately 15°C/min. And also we have achieved enough temperature uniformity during cooling even without the installation of a stirring fan in this furnace.

The temperature of the pressure vessel inner surface was raised to 120°C, and then stabilized at about 110°C. This clearly demonstrates the benefits of the new rapid cooling method – "prevention of excessive temperature rise at the pressure vessel inner surface, and best use of heat extraction capability of the pressure vessel." The design temperature of the pressure vessel inner surface is 150°C that allows for the heat extraction capability of the pressure vessel. Therefore, faster cooling can be achieved by increasing the rapid cooling gas flow rate.

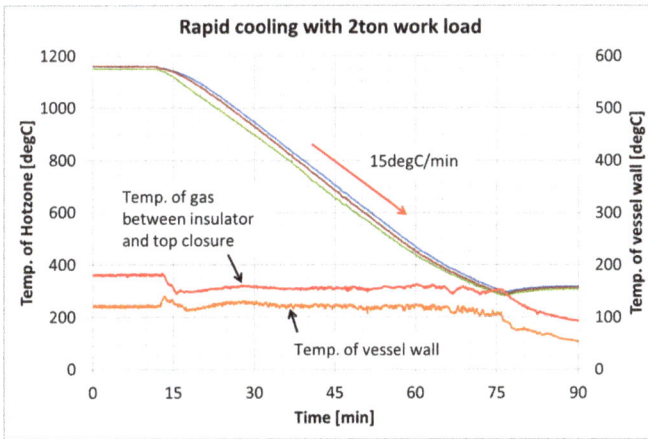

*Fig. 4. Results of the rapid cooling controlled in 15°C/min*

## Other topics

A Tungsten heater is now under development. With a Tungsten heater it is possible to have a cleaner HIP atmosphere at higher temperatures. This will increase the possibility of developing new materials. For example, materials for electronic components, certain kinds of ceramics which need contamination control, etc.

## Summary

KOBELCO has developed a new rapid cooling system as previously described. But some other related developments are in progress. With these results we expect they will expand the HIP field. We sincerely hope that our HIP technologies will contribute to the development of the global industry.

## References

[1] US Patent 4,582,681: Method and Apparatus for Hot Isostatic Pressing

[2] US Patent 7,008,210 B2: Hot Isostatic Pressing Apparatus

# Exhaust Valve Spindles for Marine Diesel Engines Manufactured by Hot Isostatic Pressing

Alberto Lapina[1,a] *, Harro Andreas Hoeg[1,b], Jakob Knudsen[1], Tomas Berglund[2], Rune Møller[3], Jesper Henri Hattel[3]

[1]MAN Diesel & Turbo, Teglholmsgade 41, 2450 Copenhagen, Denmark

[2]Sandvik Powder Solutions, P.O. Box 54, SE-735 21 Surahammar Sweden

[3]Technical University of Denmark, Department of Mechanical Engineering, Building 425, 2800 Kgs. Lyngby, Denmark

[a]Alberto.Lapina@man.eu, [b]Harro.Hoeg@man.eu

**Keywords:** Hot Isostatic Pressing, Diesel Engine, Hot Corrosion, Nickel Alloys, Thermal Cycling

**Abstract.** The exhaust valve spindle is one of the most challenging components in the marine two-stroke diesel engine. It has to withstand high mechanical loads, thermal cycling, surface temperatures beyond 700 °C, and molten salt induced corrosion. Powder metallurgy gives the opportunity of improving the component using materials not applicable by welding or forging. Therefore exhaust valve spindles have been produced by Hot Isostatic Pressing (HIP) with a spindle disc coating of a Ni-Cr-Nb alloy that cannot be manufactured by welding or forging.

This paper presents the service experience gathered by MAN Diesel & Turbo in a number of service tests on ships (up to 18000 running hours): corrosion and degradation phenomena in the spindles produced by HIP are presented and compared with the performance of state-of-the-art exhaust valve spindles.

The macroscopic geometrical changes experienced by the spindles are studied by means of Finite Element Method (FEM) calculations and strategies for further development of the component are outlined.

## Introduction

The exhaust valve spindle is one of the most challenging components in the marine two-stroke diesel engine: it has to withstand high mechanical and thermal loads without benefitting from the water cooling applied to the cylinder cover and cylinder liner.

Fig. 1 shows a cross section view of the exhaust valve spindle in the closed position, i.e. when the seat area of the spindle is in contact with the bottom piece of the exhaust valve, therefore sealing the combustion chamber. The spindle bottom is directly exposed to the combustion chamber and is the hottest part of the component, with temperatures up to 600-700 °C; the seat can go up to 450-500 °C instead.

During each engine cycle the spindle is heated up during the combustion step and thereafter cooled by the gases leaving the combustion chamber: therefore the spindle disc undergoes thermal cycling for each combustion cycle.

The high temperatures experienced by the spindle bottom reduce service life of the exhaust valve spindles because of "hot corrosion", i.e. corrosion due to the formation of molten salts, such as $Na_2So_4$ and $V_2O_5$, which dissolve the protective $Cr_2O_2$ oxide layers on the Ni-Cr alloys currently used. On the bottom of the spindle corrosion rates are usually in the range of 0.1-0.4 mm/1000 hours, but can soar up to 1 mm/1000 hours in harsh conditions.

Currently two designs for exhaust valve spindles are available: forged Nimonic80A and Duraspindle™. In the first the material Nimonic80A (20 wt% Cr, 2.5 wt% Ti, 1.5wt% Al, bal. Ni) has excellent hot corrosion resistance and is precipitation hardened to reach sufficient hardness of the seat area. The Duraspindle™ design is a stainless steel forged substrate with welded Inconel 625 on the spindle bottom and Inconel 718 on the spindle seat. The Inconel 718 layer is cold deformed and aged to increase the hardness of the seat.

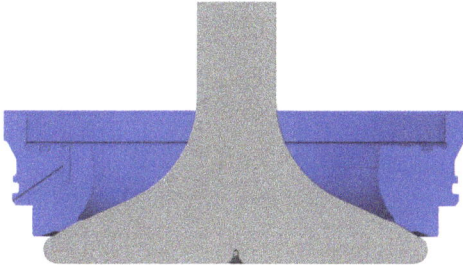

*Fig. 1 Cross section of exhaust valve spindle disc (grey) and bottom piece (blue).*

It was decided to test a hipped spindle because HIP allows a wider choice of materials for the spindle seat and bottom, without being limited by the forging and welding processes [1]. Moreover, it allows the flexibility of testing multiple materials at once by embedding samples in the spindle bottom [2]. This paper presents the service experience gained by MAN Diesel & Turbo with HIP spindles and the challenges encountered.

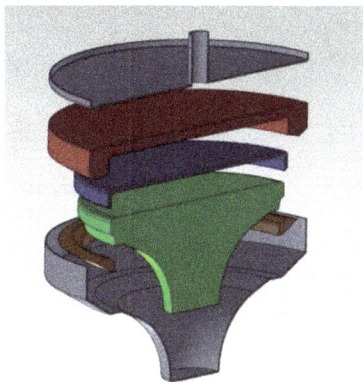

*Fig. 2 The capsule design. The grey capsule parts are produced as deep drawn steel sheet being assembled by gas tungsten arc welding (GTAW). The forged disc substrate (stainless steel) is green, the seat material orange, the bottom coating (NiCr49Nb1) red and the bonding zone material (316L) blue [1]*

Hot Isostatic Pressing – HIP'17                                          Materials Research Forum LLC
Materials Research Proceedings **10** (2019) 98-106            doi: http://dx.doi.org/10.21741/9781644900031-14

## Experimental

Manufacturing of the spindles. the spindles were manufactured at Sandvik Powder Solutions by Hot Isostatic Pressing, according to a manufacturing procedure used for a previous test [1].

Figure 2 shows an exploded view of the spindle capsule. The substrate is the same forged stainless steel used for Duraspindle. The seat material is an experimental Ni-Cr-Nb alloy. The bottom coating is NiCr49Nb1 alloy (49% wt Cr, 1.5 wt% Nb, Ni bal.). A 316L buffer layer is applied between the NiCr49Nb1 bottom coating and the forged substrate to act as a diffusion barrier to prevent the formation of chromium carbides. After the hipping at 1100 °C, the spindles undergo cold deformation and a precipitation hardening heat treatment to obtain high hardness in the seat area.

The spindles are then put in service on ships running on heavy fuel oil, for up to 18000 hours.

Finite element method (FEM) simulations. Because of the axial symmetry of the spindle, it is sufficient to have a 2D model of half the spindle disc (Fig. 3). The material used for the valve seat is the same as used for the spindle bottom: this will influence the stress distribution but only locally. The model consists of ~260000 three node axisymmetric linear elements with 130000 nodes. The model is analyzed using the Abaqus explicit solver. The three material layers are connected at the nodes, i.e. the layers share nodes. The model is held by constraining the nodes on the in the axial direction and the nodes on the centerline in the radial direction.

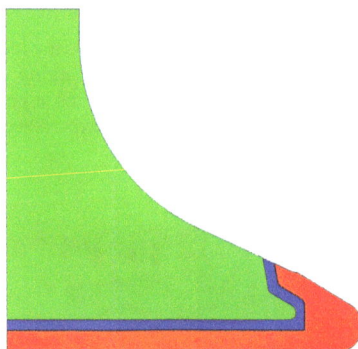

*Fig. 3 FEM model of half the spindle disc.*

The three materials involved are described with isotropic hardening and power law creep. The plastic material parameters are based on tensile tests of the materials whereas the creep parameters are based on literature data. An attempt to measure and apply measured creep parameters is ongoing.

The in-service mechanisms have been simplified so that the model experiences only a constant temperature field during the combustion, i.e. the entire spindle disc is uniformly heated up to 700 °C, without external or internal forces applied, meaning that the occurring temperature gradient across the radius of the valve spindle is neglected. The contribution of this to the bending is considered insignificant. The temperature cycling due to the combustion cycle is not considered.

The FEM analysis covers the HIP stage and up to five cycles of normal operation. The HIP stage involves uniform cool down from 1100°C to 20°C. Each operating cycle describes heating up to 700°C, holding for 260 h and then cooling to 20°C.

**Results**

Corrosion rate of the bottom coating. Because of the bending of the spindle taking place during service (which is illustrated in Section "Bending of the spindle) it is not possible to measure directly the absolute value of the corrosion rate for NiCr49Nb1. However Nimonic80A and Inconel 625 samples were embedded in the bottom coating of two of the spindles during production, therefore it is possible to measure the difference in corrosion rates between these materials and NiCr49Nb1. Both materials have higher corrosion rates than NiCr49Nb1: approximately 0.07 mm/1000 hours for Nimonic80A and 0.15 mm/1000 hours for Inconel 625.

Mechanical properties of NiCr49Nb1 and microstructure evolution. Prior to service, samples of NiCr49Nb1 have been heat treated at 700 °C for up to 4400 hours, in order to simulate the effect of service conditions on the material. Fig, 4 shows the mechanical properties of NiCr49Nb1 in the as-hipped conditions and after different heat treatment times.

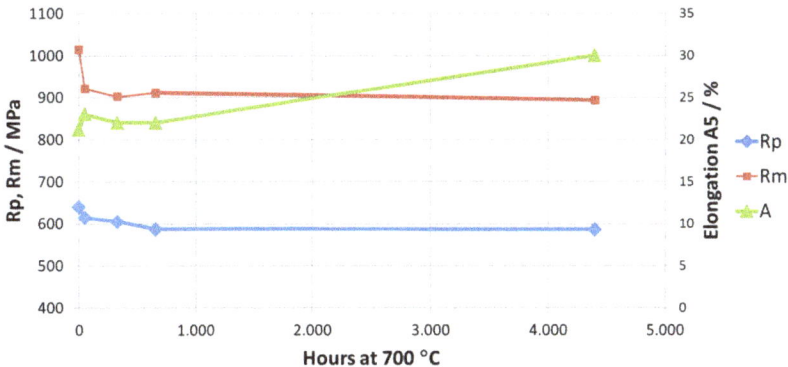

*Fig. 4 Mechanical properties of NiCr49Nb1 as a function of time at 700 °C.*

Yield strength and tensile strength decrease significantly within the first hundred hours at 700 °C and then stay constant over time. Ductility instead increases significantly over a thousands-of-hours timescale.

The evolution of the material microstructure is investigated by electron microscopy: a significant change in phase distribution is seen after heat treatment at 700 °C: Fig. 5 shows backscattered electron images of as-hipped material and material after heat treatment at 700 °C. The fraction of α-Cr increases and a Nb rich phase (the bright phase in the electron micrographs) appears. The same has been observed in NiCr49Nb1 after service test.

Hot Isostatic Pressing – HIP'17                                    Materials Research Forum LLC
Materials Research Proceedings **10** (2019) 98-106          doi: http://dx.doi.org/10.21741/9781644900031-14

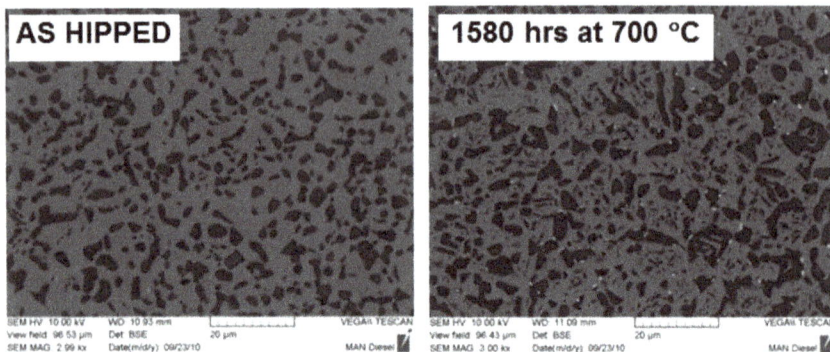

*Fig. 5 Backscattered electron images of NiCr49Nb1as-hipped and after 1580 hours at 700 °C.*

Bending of the spindle. During service the spindle disc bends upwards, as sketched in Fig. 6. This causes the seat angle (defined as the angle between the seat surface and the direction perpendicular to the spindle stem axis, see Fig. 6) to decrease during service from the starting value of 30.3° degrees.

The change in seat angle is plotted versus the service time in Fig. 7: the seat angle continues to change even after 10000 hours in service.

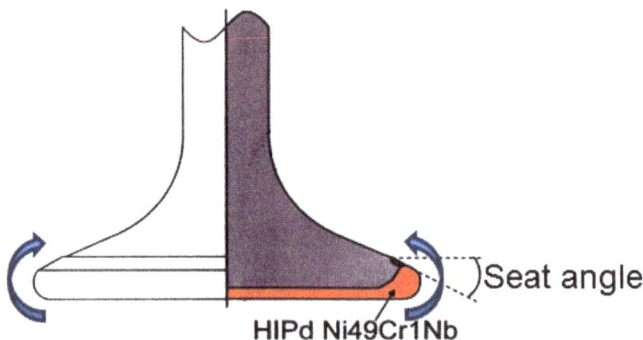

*Fig. 6 Sketch of the spindle showing the seat angle and the direction of the bending.*

Crack formation at the seat. Penetrant testing reveals that all the spindles, after service, have a crack running along the inner circumference of the seat area. Fig. 8 shows a polished cross section of the seat area after etching with Marble solution. The crack is perpendicular to the surface, about 3 mm long and crosses almost entirely the 316L buffer layer.

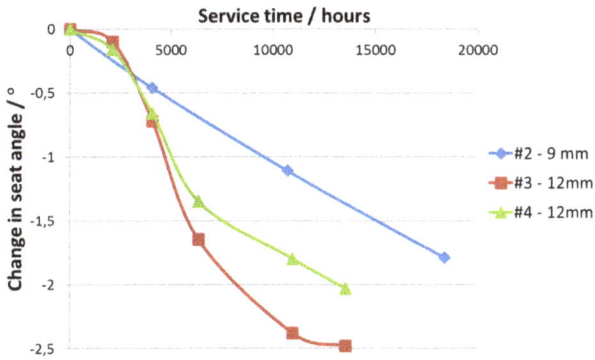

*Fig. 7 Change of seat angle over service time for spindles 2, 3 and 4. "9/12 mm" refers to the thickness of the bottom NiCr49Nb1 coating.*

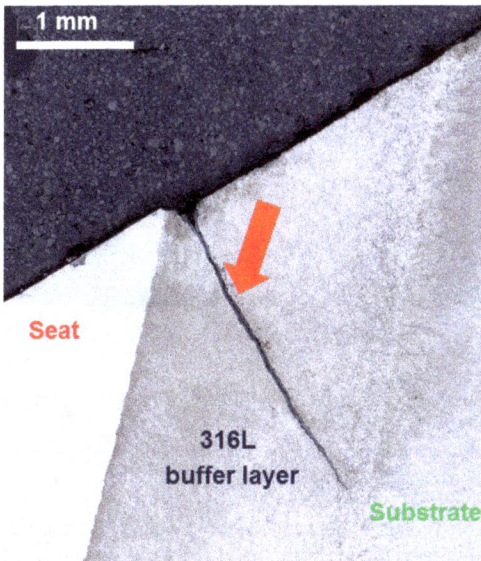

*Fig. 8 Optical microscopy of cross section of the seat area after etching.*

FEM calculations. Fig. 9 shows Von Mises stresses at room temperature, after the first thermal cycle and the fifth thermal cycle. Note that the resulting deformation is exaggerated/amplified to better show the bending. The von Mises stresses increase slowly with the number of cycles. The results show high stresses in the seat region and stress concentrations at the outward corners of the 316L layer. Concerning radial stresses, the NiCr49Nb1 coating experiences compressive stresses at the center of the spindle bottom, while the 316L layer is experiencing tensile stresses, see Fig. 10. This induces a bending moment forcing the nose upwards. Yielding is predicted to occur in the 316L layer and the seat area.

The seat angle is calculated after each of the five thermal cycles. The angle continues to decrease (i.e. the spindle bends upwards) at each cycle, reaching a change of -0.068° after 5 cycles.

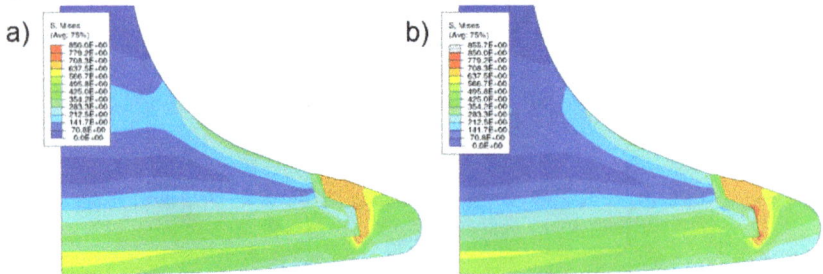

*Fig. 9 Von Mises stresses after a) the 1$^{st}$ and b) 5$^{th}$ service cycle.*

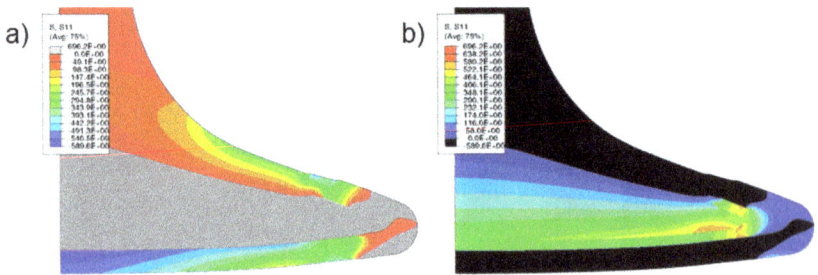

*FIG.*

*10 a) Compressive and b) tensile radial stresses after the after the 5$^{th}$ service cycle.*

## Discussion

Corrosion resistance. In this service test NiCr49Nb1offers only a minor improvement when compared to Nimonic80A and Inconel 625.

Previous service tests with hipped spindles on a 900 mm cylinder-bore engine showed instead higher differences in corrosion resistance (up to 0.3 mm/1000 hours for Nimonic80A and 0.57 mm/ 1000 hours for Inconel 625).

Such difference in corrosion rates between the current and previous service tests is most likely due to a higher operating temperature in the previous test. It is known that hot corrosion rates follow an Arrhenius rate law [4], thus increasing with increasing temperature.

Mechanical properties of NiCr49Nb1 and microstructure evolution. The changes in microstructure in NiCr49Nb1 can be understood on the basis of the JMatPro simulation of phase composition at equilibrium provided in Fig. 11.

Hot Isostatic Pressing – HIP'17                                    Materials Research Forum LLC
Materials Research Proceedings **10** (2019) 98-106          doi: http://dx.doi.org/10.21741/9781644900031-14

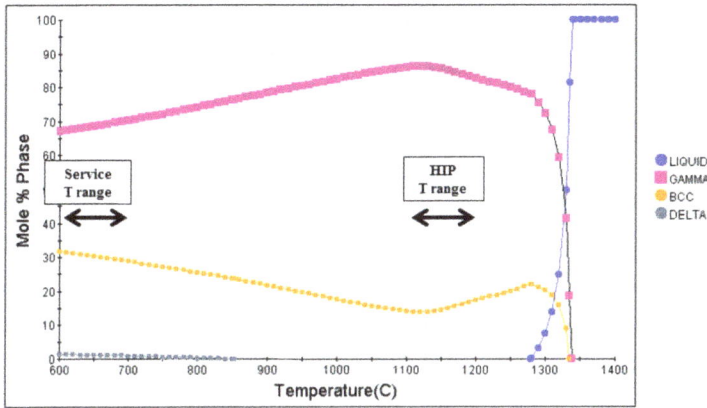

*Fig 11 Simulation of phase composition of NiCr49Nb1 as a function of temperature.*

There is significant difference in equilibrium phase distribution between the hipping temperature (1100 °C) and the service temperature (600-700 °C). A service temperature of 600-700 °C is sufficiently high to allow diffusion processes to take place and change the phase distribution toward equilibrium conditions, increasing the fraction of α-Cr and forming Ni$_3$Nb particles. This behavior is observed in the material (see Fig. 5) and has been seen previously in a service test carried out on a 900 mm cylinder-bore engine.

The changes in microstructure are the reason for the changes in mechanical properties after heat treatment at 700 °C.

Bending of the spindle and FEM calculations. During service the spindle disc deforms in-elastically resulting so that the edge bends upward. This type of deformation is unique to the HIP spindle and is not observed with the current design. The root cause must therefore be found in the specific materials and processes used for the HIP spindle.

Dimensional considerations show that the deformation cannot be caused by the high-cycle thermal fatigue of the combustion cycle: the observed deflection, divided by the number of thermal cycles, would give a step value smaller than the radius of a hydrogen atom, therefore too small to be plausible.

The simulations indicate that the deformation is caused by differences in material properties of the materials involved. When the spindle heats up the base material and the 316L buffer layer expands inducing compressive stresses in the buffer layer and tensile stresses in the corrosion layer. During operation the 316L layer relaxes due to creep. When cooling down the base material and buffer layer contract inducing compressive stresses in the NiCr49Nb1 anti corrosion layer [5].

The simulation shows that the stresses in the NiCr49Nb1 coating of the seat region are so high that yielding under compression is expected. In the real spindle however, the seat material has higher yield stress than in the model. Furthermore, the simulations indicate high stress concentrations at the sharp outward corners of the 316L layer. The stress concentration is expected to be less pronounced in the real spindles since some mixing of the powders is unavoidable, reducing the metallurgical notch. It is not possible to model material gradients in a FEM model.

Crack at the seat. As for the bending of the spindle, the crack at the inner circumference of the seat is a damage feature unique to the HIP spindle, never observed with the current design.

The "step" (indicated by the red arrow in Fig. 8) at the seat/buffer layer interface is likely due to the different corrosion resistance of NiCr40Nb3,5 and 316L: after machining the surface is flat across the material interface, but 316L corrodes more rapidly than the seat material and so a "step" at the material interface is formed.

This step formation might likely contribute to stress concentration acting as a notch and therefore make the material in this area more prone to cracking.

Moreover, the increase in seat angle over service time means that the contact point with the bottom piece moves toward the outer part of the seat area, and this increases the momentum and therefore the stresses in the inner seat area.

## Conclusions and future work

The service tests have proven that a hipped spindle with NiCr49Nb1 can operate successfully for thousands of hours. The powder metallurgy route allows using material previously not available because of welding and forging limitations.

NiCr49Nb1 has better corrosion resistance than the current state of the art materials for the spindle of bottom: the level of improvement depends on the engine and running conditions.

After an initial drop, its strength is stable over thousands of hours at service temperature.

The major challenges faced by this new design are the formation of cracks at the inner seat circumference and the bending of the spindle.

The first issue is likely caused by the second one; therefore future work will concentrate on preventing the spindle bending.

In this respect a change in the design of the buffer layer (e.g. reduced thickness, change in shape) is currently considered, and other materials are under testing.

Future simulation work will include temperature field in the spindle, and once experimental creep data for the materials involved are available, they will be implemented in the FEM model.

## Acknowledgments

This development project has been supported by EUDP (Energy Technology Development and Demonstration Program), funded by the Danish government.

We thank Leonhardt & Blumberg and Wallenius Marine for allowing service testing of the exhaust valve spindles onboard their ships.

## References

[1] U.D. Bihlet, H. a Hoeg, Future HFO / GI exhaust valve spindle, in: CIMAC, Shangai, 2013.

[2] U. Bihlet, H. Hoeg, K.V. Dahl, M.A.J. Somers, In-situ hot corrosion testing of candidate materials for exhaust valve spindles, in: Int. Conf. Hot Isostatic Press., 2011.

[3] R. Møller, FE Analysis of Temperatures and Stresses In a Spindle During In-Service Conditions, 2016.

[4] J.R. Nicholls, D.J. Stephenson, Hot corrosion tests on candidate diesel valve materials, in: Diesel Engine Combust. Chamb. Mater. Heavy Fuel Oper., 1990: pp. 47–60.

[5] R. Møller, Residual stresses and in-service creep of a sintered multi-material component, Technical University of Denmark, 2014.

Hot Isostatic Pressing – HIP'17
Materials Research Proceedings **10** (2019) 107-113

Materials Research Forum LLC
doi: http://dx.doi.org/10.21741/9781644900031-15

# Microstructural Design of Ni-base Superalloys by Hot Isostatic Pressing

Benjamin Ruttert [1,a *], Inmaculada Lopez-Galilea [1], Lais Mujica Roncery [2] and Werner Theisen [1]

[1] Lehrstuhl Werkstofftechnik, Ruhr-Universität Bochum, 44801 Bochum, Germany

[2] Grupo de Investigación en Materiales Siderúrgicos e INCITEMA, Universidad Pedagógica y Tecnológica de Colombia, Boyacá, Colombia

[a] benjamin.ruttert@.rub.de

**Keywords:** HIP, Microstructure, Ni-base Superalloy, Porosity, Rejuvenation

**Abstract.** Single-crystal Ni-base superalloys (SXs) are used as a first-stage blade material in high-pressure turbines for aero engines or in stationary gas turbines. They operate at temperatures close to their melting point where they have to withstand mechanical and chemical degradation. Casting and extensive solution heat-treatments of such blades introduce porosity that can only be reduced by hot isostatic pressing (HIP). Recent developments in HIP plant technology enable simultaneous HIP-heat-treatments due to rapid quenching at the end of such treatments. This work gives an overview of the opportunities that such a unique HIP offers for the solution heat-treatment of conventionally cast SXs or directionally solidified Ni-base superalloys fabricated by selective electron beam melting (SEBM). The influence of temperature, pressure, and cooling method on the evolution of the $\gamma/\gamma'$-morphology and on the pore shrinkage is investigated. The cooling method has a strong impact on the $\gamma'$-particle size and shape whereas the combination of temperature and pressure during the HIP-treatment mainly influences porosity reduction. In a final approach a HIP treatment is satisfactorily used to fully re-establish the $\gamma/\gamma'$-microstructure after high-temperature creep degradation.

## Introduction

SXs are used as a first-stage blade material in modern gas turbines [1]. Their complex composition results in large dendrite arm spacings during the slow Bridgman solidification process, with heavy partitioning of alloy elements between dendritic and interdendritic regions as well as the formation of large cast pores in interdendritic regions. The presence of porosity reduces the material strength and ductility and results in scattering of the mechanical properties. Pores act as crack initiation sites and promote crack propagation, leading to premature rupture of the components [2-3]. Therefore, it is important not only to reduce the segregation by a heat-treatment (solution annealing and aging), but also to reduce the porosity generated during casting and solution annealing by means of HIP. Modern HIP units can provide fast quenching rates that help in designing the desired microstructures starting from material states that feature internal pores, undesirable precipitates, and chemical segregation. The simultaneous application of a high isostatic pressure and a high temperature can eliminate pores by a combination of elementary processes that involve plastic deformation, creep, and diffusion bonding and also simultaneously remove chemical heterogeneities of the alloy. The possibility of controlling the cooling rate after HIP to a certain degree (from quenching to slow cooling) enables establishment of a desired final $\gamma/\gamma'$ microstructure at the end of such an implemented HIP-heat-treatment [4]. Consequently, the combination of HIP and quenching enables integration of the required homogenization of the superalloys within the HIP-process, thus resulting in one processing step [5]. The present contribution intends to give an overview of the influence of HIP parameters such as temperature, pressure, and cooling rate of modern HIP units on porosity reduction as well as on the evolution of the $\gamma/\gamma$'-microstructure of cast SX as wells as additively manufactured Ni-base superalloys. Since SX turbine blades operate under harsh service conditions (high temperatures and stresses), time-dependent microstructural changes occur that degrade the microstructure and thus the lifetime of the blades: namely rafting and the formation and growth of cavities. The high costs of SX components has led to an increased interest in extending their service live by various repair and rejuvenation procedures [6]. Lastly, a short approach is made in this work to rejuvenate the crept microstructure by an appropriate HIP-rejuvenation.

**Materials and methods**
In this work, the ERBO/1 SX is investigated. It is a CMSX-4 type of alloy with a specific heat-treatment [7]. ERBO/1 is used in three different states: as-cast (ERBO/1A), after solution annealing in a laboratory vacuum furnace (ERBO/1B) and after conventional solutioning and subsequent precipitation hardening (ERBO/1C). All details regarding the chemical composition, homogeneity, and microstructural details have been described elsewhere [7]. All specimens used in this work were precisely oriented in the <100> direction by combining the Laue technique with electro discharge machining [7]. The specimens for scanning electron microscopy (SEM) were examined perpendicularly to the solidification direction for microstructural characterization at the dendrite core and parallel to the solidification direction for porosity quantification. For the porosity measurement, SEM back-scattered electron panorama montages (magnification: 500x), covering a total area of 3 mm x 3 mm were taken. The porosity was determined in the (100)-plane, parallel to the dendrite growth direction. The healing of pores is described, by determining the total measured pore area of ERBO/1(A, B or C) divided by the remaining pores after the different HIP treatments. Quantitative microstructural analysis of the porosity and the $\gamma/\gamma$'-microstructure was supported by the software Image J. High-resolution dilatometry was carried out to determine the $\gamma'$-solvus temperature [5]. HIP was performed in two different HIP facilities. The first one of type QIH-9 URQ, from Quintus Technologies, allows ultra-rapid quenching (up to 2000 K/min), as well as very low cooling rates with controlled cooling conditions. The second one of type QIH-9, is able to reach cooling rates of up to 200 K/min. All HIP experiments were carried out in molybdenum furnaces under an inert Ar atmosphere.

**Results and discussion**
The main characteristic parameters of an as-processed SX microstructure are its $\gamma/\gamma$'-phase morphology and the cast porosity. The temperature and hydrostatic pressure parameters of HIP govern the kinetics of pore shrinkage during the process. However, the cooling rate in combination with the HIP temperature down to room temperature govern the evolution of the $\gamma/\gamma$' microstructure. In order to reduce porosity by a HIP treatment, it is important to apply a temperature that is higher than the $\gamma$'-solvus temperature because pressure-driven material flow is fast only if the soft $\gamma$-phase is present. At temperatures below $T_{\gamma'\text{-solvus}}$, $\gamma$'-particles are present that strengthen the $\gamma$-matrix. The resulting increase in creep resistance makes compaction associated with pore shrinkage more difficult. Fig. 1 shows how the HIP temperature and HIP pressure affect the healing of pores. The porosity values of the homogenized material in the laboratory ERBO/1B (without HIP) are 0.365 area% pores, 48 pores/mm$^2$ and 10 µm average pore diameter [5].

Materials Research Forum LLC
doi: http://dx.doi.org/10.21741/9781644900031-15

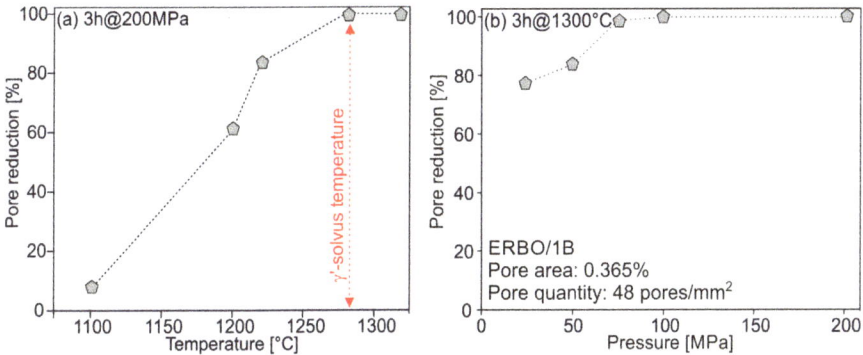

*Fig. 1. Influence of HIP parameters on pore healing. (a) Effect of HIP temperature for 3h HIP exposure and a pressure of 200 MPa on conventionally heat-treated ERBO/1C. (b) Effect of HIP pressure at a temperature of 1300°C and 3h exposure on homogenized ERBO/1B. The porosity values of conventionally heat-treated ERBO/1B are given as a reference. Figure modified from ref. [5]*

Fig. 1a shows the results of HIP experiments in which a HIP pressure of 200 MPa was applied for 3 hours at different HIP temperatures. The $T_{\gamma'\text{-solvus}}$ was determined to be 1285°C using a high-resolution calorimetry method (shown as a dashed vertical red line in Fig. 1a) [5]. Fig. 1a shows that full porosity reduction can only be achieved at temperatures above the $T_{\gamma'\text{-solvus}}$ and also shows that 1100°C is not a sufficiently high HIP temperature for eliminating as-cast porosity. Higher temperatures favor pore shrinkage because material flow due to high-temperature plasticity is faster (mechanical aspect) and because the diffusion coefficients of alloy elements increase (kinetic aspect). Fid. 1b shows the effect of HIP pressure on porosity reduction for HIP treatments performed at 1300°C for three hours and also shows that porosity reduction becomes more effective when the HIP pressure increases up to 75 MPa, taking the overall pore area of ERBO/1B into account. However, from this pressure value onwards, a further pressure increase does not further accelerate pore healing. It is interesting to note that even low HIP pressures such as 25 and 50 MPa achieve porosity reductions of 77 and 99%, respectively. The fact that higher pressures result in more effective porosity reduction results from the fact that higher pressures represent higher driving forces for plasticity controlled compaction. Typical porosity values after HIPing at 1300°C for 3h at 100 MPa pressure are 0.001 area% pores, 1 pores/mm$^2$ and 2.6 μm average pore diameter [5]. The features of the γ/γ' microstructure strongly depend on the cooling rate after the isothermal HIP treatment. This has been studied for the alloy ERBO/1B. Different cooling rates were applied after 3h isothermal HIP treatments at 100 MPa and 1300°C. After HIP, three different cooling rates were applied: fast, intermediate, and slow. By approximating the cooling curves as straight lines in the temperature interval between 1300 and 800°C, these three cooling rates can be approximated as 30 K/s (fast), 1 K/s (intermediate) and 0.3 K/s (slow). It was found that the cooling rates after HIP did not affect the porosity. However, they have a strong influence on the γ/γ'-microstructure, as shown in Fig. 2a to 2c. Decreasing cooling rates result in an increasing particle size and decreasing particle number fractions. Faster cooling rates are associated with smaller particles in which smaller elastic strain energies are less effective in enforcing cuboidal shapes. Whereas the effect of cooling rate on the particle structure directly after cooling is more or less as expected, it is worth noticing that different cooling rates effect the γ/γ' microstructures, even when two-stage post-HIP heat-treatments of 1140°C for 4h and 870°C for 16h at atmospheric pressure are applied (see Fig. 2d to 2f). The results shown in Fig. 2d to 2f clarify that these additional heat-treatments result in a controlled type of particle coarsening which establishes the final microstructure. When the systems starts with a larger γ' particle size, it also has larger γ'-particle sizes after the additional heat-treatment steps, as can be seen by comparing Fig. 2d and 2f. Post-HIP annealing heat-treatments also result in a higher degree of regularity of the γ/γ'-microstructure.

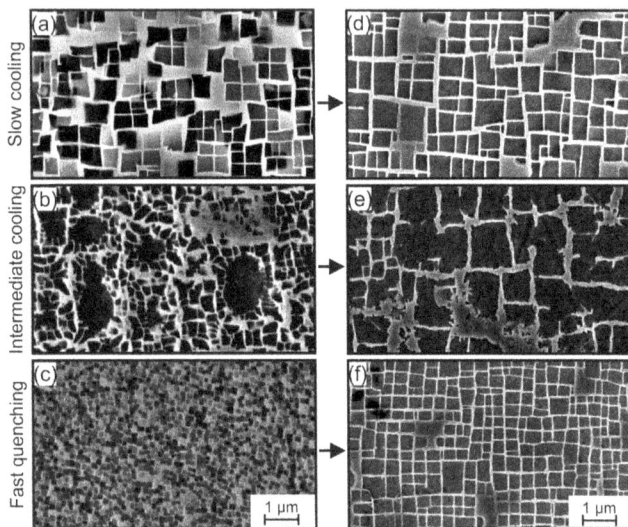

*Fig. 2. SEM micrographs illustrating the effect of the cooling rate on the microstructure that terminates the HIP cycle at 100 MPa pressure. (a)-(c) Microstructures after different cooling rates. (d)-(f) Microstructures after cooling and an additional two step aging.*

Electron-beam melting was used to build cylindrical, columnar-grained parts from pre-alloyed and atomized CMSX-4 powder [8]. The microstructure of such parts is very fine and the dendrite arm spacing is several orders of magnitude finer than that of cast parts. Therefore, the required homogenization time of such parts to reduce elemental partitioning can be significantly shortened. A first approach was made in this work, in which homogenization of a SEBM part was transferred into a novel HIP with quenching capability (see Fig. 3a). SEBM parts in the as-built condition still exhibit microstructural heterogeneities such as small pores (Fig. 3b) and small-scale elemental partitioning as shown qualitatively in the Re-element mapping in Fig. 3c. The results of Fig. 3d and 3e demonstrate that only a few minutes of isothermal holding time in combination with a moderate heating rate are sufficient to almost completely shrink the porosity (Fig. 3d) and to dissolve the elemental partitioning (Fig.3e). Rapid quenching at the end of the holding step leads to small and uniformly distributed γ'–precipitates, as described previously in this section. Additionally, due to the layerwise manufacturing during SEBM (in-situ heat-treatment), the resulting γ'–precipitate size is a function of the distance from the bottom to the top surface of the manufactured parts [8]. At the HIP-homogenization temperature, however, the γ'-size gradient is dissolved first and subsequent rapid quenching produces a fine and uniform γ'-size distribution that is favorable for the following precipitation hardening.

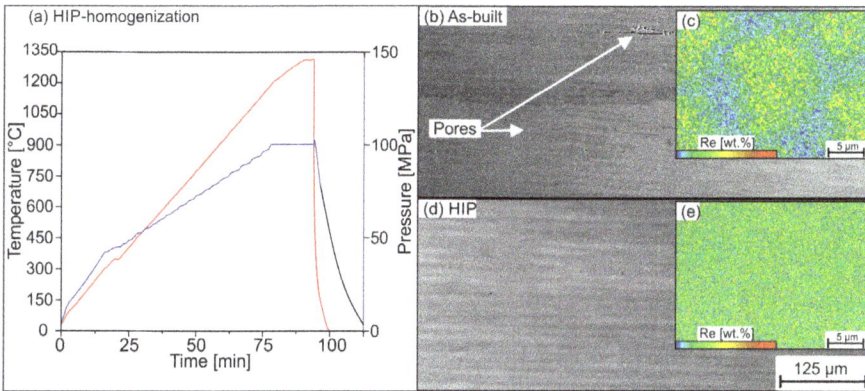

*Fig. 3. HIP heat-treatment for the EBM part and SEM micrographs showing the microstructural evolution. (a) HIP treatment at a pressure of 100 MPa and 30 min HIP exposure. (b) Microstructure in the as-built longitudinal section, showing pores, and (c) showing an EDS element mapping for Re. (d) and (e) show the corresponding microstructures after the HIP treatment.*

At least a first step has been made to explore whether the γ/γ'-microstructure of the initial material state can be re-established by HIP after high-temperature creep deformation. A conventionally heat-treated ERBO/1C specimen was creep deformed at 1050°C and 160 MPa to a final strain of 5%. The microstructure of this material state was investigated in the SEM. Fig. 4a shows an SEM micrograph of the rafted microstructure after high-temperature creep deformation. The specimen was then subjected to the HIP-rejuvenation treatment shown in Fig. 4b. After 4h of isothermal HIP with rapid quenching (fast cooling, section I), subsequent precipitation hardening (4 hours at 1140°C and 16 hours at 870°C) was performed in a laboratory furnace under an Ar atmosphere at atmospheric pressure. The microstructural evolution of the crept and subsequently rejuvenated material states are compared in Fig. 4a and 4c, respectively. It can be clearly seen that the rafts have disappeared and that a fine and uniform γ/γ'-microstructure was re-established. Further work is required to fully exploit the potential of a HIP rejuvenation treatment. The effect of rejuvenation on porosity, dislocations, recrystallization, on the presence/absence of TCP phases, and lastly on additional creep life must be investigated carefully.

## Summary and conclusions
The present work investigates the effect of a novel HIP treatment on the microstructure of an as-cast / as-built, as-cast and heat-treated, and a pre-crept SX of type ERBO 1 (CMSX 4 family). It was shown that HIP treatments can heal porosity. In order to optimize HIP-parameters, temperatures must be higher than the γ'-solvus temperature and pressures consequently higher

than 75 MPa. However, the material's porosity is not affected by the cooling rates after isothermal HIP treatments. In contrast, the cooling rates have a strong influence on the $\gamma/\gamma'$-microstructure. Faster cooling rates result in finer microstructures, directly after HIPing as well as after a subsequent precipitation hardening. A HIP rejuvenation treatment has been shown to restore a fine and regular $\gamma/\gamma'$-microstructure after initial high-temperature creep deformation.

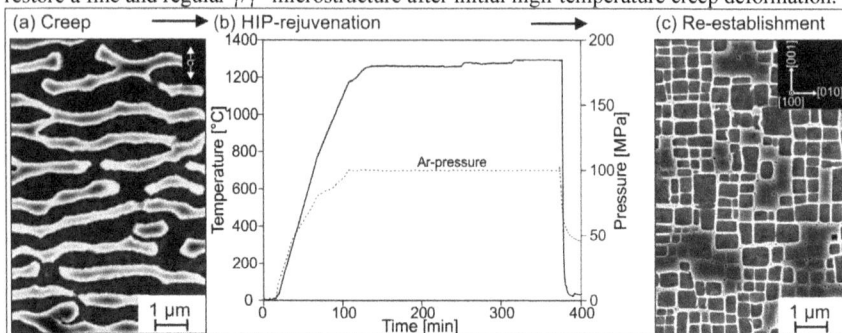

*Fig. 4. SEM micrographs showing the effect of HIP-rejuvenation on the microstructure after creep. (a) Topological inversion, and (b) temperature vs time and pressure vs time histories during the rejuvenation. At the end of isothermal HIP exposure: fast quenching. (c) Re-established microstructure.*

**Acknowledgements**

The authors acknowledge funding by the Deutsche Forschungsgemeinschaft (DFG) via the collaborative research center SFB/TR-103 through project B4 (BR, ILG, LMR, WT). Assistance from D. Bürger, P. Wollgramm and G. Eggeler from project A1 for providing the creep specimen and M. Ramsperger and C. Körner from project B2 for providing the EBM parts is gratefully acknowledged.

**References**

[1] G.W. Meetham, The Development of Gas Turbine Materials, UK, London, (1981). https://doi.org/10.1007/978-94-009-8111-9

[2] A. Epishin, T. Link, Mechanisms of high-temperature creep of nickel-based superalloys under low applied stresses, Philos. Mag. A. 84 (2004) 1979-2000. https://doi.org/10.1080/14786430410001663240

[3] G. Mälzer, R.W. Hayes, T. Mack, G. Eggeler, Miniature Specimen Assessment of Creep of the Single-Crystal Superalloy LEK 94 in the 1000°C Temperature Range, Metall. Mater. Trans. A 38 A (2007) 314-327.

[4] H.V. Atkinson, B.A. Rickinson, Hot Isostatic Processing, UK, Bristol, (1991).

[5] L. Mujica Roncery, I. Lopez-Galilea, B. Ruttert, S. Huth, W. Theisen, Influence of temperature, pressure, and cooling rate during hot isostatic pressing on the microstructure of an SX Ni-base superalloy, Mater. Des. 97 (2016) 544-552. https://doi.org/10.1016/j.matdes.2016.02.051

[6] A. Baldan, Rejuvenation procedures to recover creep properties of nickel-base superalloys by heat treatment and hot isostatic pressing techniques, J. Mater. Sci. 26 (1991) 3409-3421. https://doi.org/10.1007/BF00557126

[7] A.B. Parsa, P. Wollgramm, H. Buck, C. Somsen, A. Kostka, I. Povstugar,P. Choi, D. Raabe, A. Dlouhy, J. Müller, E. Spiecker, K. Demtröder, J. Schreuer, K. Neuking, G. Eggeler, Advanced Scale Bridging Microstructure Analysis of Single Crystal Ni-Base Superalloys; Adv. Eng. Mater. 17 (2015) 216-230. https://doi.org/10.1002/adem.201400136

[8] M. Ramsperger, R.F. Singer, C. Körner, Microstructure of the Nickel-Base Superalloy CMSX-4 Fabricated by Selective Electron Beam Melting, Metall. Mater. Trans. A 47 (2016) 1469–1480. https://doi.org/10.1007/s11661-015-3300-y

Hot Isostatic Pressing – HIP'17                    Materials Research Forum LLC
Materials Research Proceedings **10** (2019) 114-120    doi: http://dx.doi.org/10.21741/9781644900031-16

# The Effect of HIP Treatment on the Mechanical Properties of Titanium Aluminide Additive Manufactured by EBM

Daisuke Kondo[1,a*], Hiroyuki Yasuda[2], Takayoshi Nakano[2],
Ken Cho[2], Ayako Ikeda[3], Minoru Ueda[1], Yuto Nagamachi[1]

[1]Metal Technology Co Ltd., 713, Aza-narihira, Shake, Ebina, Kanagawa, 243-0424, Japan

[2]Osaka University, 2-1, Yamadaoka, Suita, Osaka, 565-0871, Japan

[3]National Institute for Materials Science, 1-2-1, Sengen,Tsukuba, Ibaraki, 305-0047, Japan

[a] d_kondo@kinzoku.co.jp

**Keywords:** HIP, Electron Beam Melting (EBM), Additive Manufacturing, Titanium Aluminide

**Abstract.** Titanium aluminide, one of the important next generation high temperature materials, attracts intense R&D interests, and the application for the aeronautics and space fields is being intensely investigated. TiAl components additively manufactured by us possesses more than 99% density and good mechanical properties, however residual voids are problematic in the area where cyclic properties are important, therefore hot isostatic pressing (HIP) treatment is necessary. In this study, the effect of HIP treatment on the lamellar structure of TiAl alloy which showed excellent tensile ductility is investigated.

## 1. Introduction

Recently, in the transport industry, especially in the aeronautic section, high efficiency becomes vitally important in regards to reducing the environmental burden, fuel cost and decreasing the amount of exhaust gases. For example, the Boeing 787 successfully reduced 20% of its fuel cost, mainly due to this increased efficiency of its engine. This increased efficiency of jet engine stems from reducing the total engine weight. One way this is achieved is through the employment of TiAl alloy with its high strength to weight ratio and high elevated temperature strength which is the reason its application has widely spread [1]. Currently, TiAl alloy is employed in low pressure turbine blades, however if higher reliability is achieved, wider use of this alloy is possible, and thus even higher efficiency will be attainable.

Conventionally, TiAl parts are manufactured by precision casting, such as the lost wax method [2]. However, TiAl alloy has inherently low fluidity when casting, which sometimes results in casting defects. Usually, turbine blades design incorporates complicated cooling channels, requiring a sophisticated manufacturing process employing various cores. Since TiAl alloys are very reactive, it has a tendency to react with the casting mold as well as forming imperfections of surface oxide. Therefore, after casting, surface machining is essential and results in considerable material loss.

This has led us to focus on an additive manufacturing method as a more efficient process, especially through using an electron beam melting 3D printing (3D EBM), because it operates in a high vacuum, thus enabling little surface contamination and a negligible amount of material loss. Furthermore, it has been reported for Co-Cr-Mo and Ti-6Al-4V alloys [3-5], that by carefully choosing operational parameters, it is possible to control not only the shape but also the microstructure and mechanical properties of parts manufactured by 3D EBM. We have started our development work on low pressure turbine blades for aeronautic jet engines, focusing on

Hot Isostatic Pressing – HIP'17                                   Materials Research Forum LLC
Materials Research Proceedings **10** (2019) 114-120        doi: http://dx.doi.org/10.21741/9781644900031-16

shape and microstructural control. By using the 3D EBM and applying optimum parameters, it was possible to manufacture TiAl parts with a microstructure consisting of layered microstructure of soft and ductile equiaxed γ grain layers (γ bands) and a hard duplex-like region as shown in Fig. 1. Fig. 2 shows that EBM-TiAl exhibits yield strength of 550 MPa or higher at room temperature regardless of orientation with respect to its aligned structure [6]. However, ductility at room temperature is strongly dependent on the layered microstructure, and when the test direction is 45 degrees to the layered microstructure, the maximum elongation of more than 2.5% is obtained, as shown in Fig. 3 [6]. The high hardness and elongation obtained in this study are the deformation concentration effect on soft γ band compared to duplex-like region and the physical properties obtained from layered structure of different properties with high strength of microstructure duplex-like region. This high room temperature elongation is derived from the existence of γ bands which play a key role in the shear deformation to the maximum shear direction. Even though the 3D EBM results in a high density of 99% or higher, it is still prone to defects or porosities [7].

Fig. 1 Microstructure of EBM-TiAl [6]

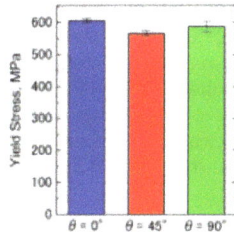

Fig. 2 Yield stress of EBM-TiAl (room temperature) [6]

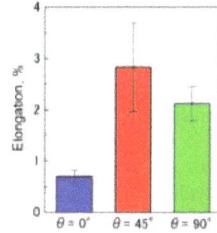

Fig. 3 Elongation of EBM-TiAl (room . [6]

Our previous study on a 3D EBM manufactured Ti6Al4V alloy revealed the existence of porosities in the as-build condition, and by applying hot isostatic pressing (HIP) treatment, fatigue properties of the alloy can be drastically improved as a result of pore elimination [8]. Since the low pressure turbine blades are used in a harsh environment, it is regarded as important to have a high reliability for EBM-TiAl through HIP treatment.

In this study, we sought to reduce the defects and porosities found in EBM-TiAl while maintaining the characteristic layered microstructure, and investigated the optimum HIP conditions whilst focusing on the microstructure. In addition we also investigated the influence of the HIP conditions on the mechanical properties of the EBM-TiAl.

## 2. Experimental Procedure

The powder of the Ti-48Al-2Cr-2Nb (at %) alloy, used in the present study, consists of spherical particles of which average diameter is approximately 100µm. TiAl cylindrical rods of 10 mm diameter and 90 mm length were built using an Arcam A2X$^{TM}$ using Ti-48Al-2Cr-2Nb powder; the composition is shown in Table 1. Details of manufacturing conditions are made in previous reports. [6]

*Table 1 Chemical composition of TiAl alloy powder (at.%)*

|        | Al   | Cr   | Nb   | C     | O     | N     | Ti   |
|--------|------|------|------|-------|-------|-------|------|
| Powder | 48.6 | 1.74 | 1.95 | 0.032 | 0.198 | 0.008 | Bal. |

In this study, direction is defined as 0 degrees when the building direction is parallel to the bar axis. HIP treatments at 1100 °C and 1250 °C for 3 hours were performed on the as-built material. The microstructures of the alloy were examined with an optical and scanning electron microscope before and after HIP treatment. The specimens for the observation were electrically polished in a $HClO_4$ : butanol : methanol (6 : 35 : 59 vol%) solution.

Defects were observed by optical microscopy over 550 $mm^2$ area, while pores were investigated by scanning electron microscopy over 0.05 $mm^2$ area. The number of samples for each defect and pore observations is not enough for determining appropriate standard deviation.

The volume fraction of $\alpha_2$ phase in duplex-like region was determined by counting area percent of $\alpha_2$ phase under scanning electron microscope over 0.2$mm^2$ area. The resultant area percent was used for the volume fraction.

Mechanical properties of the sample were determined with tensile and fatigue tests conducted at RT. Specimen shapes for both the tensile test and fatigue tests are shown in Fig. 4 and both the tensile and fatigue specimens were polished with colloidal $SiO_2$ suspension. For the tensile test, an Instron-type testing machine was used, and strain rate was chosen at $1.7 \times 10^{-4}$ $s^{-1}$. The fatigue test was performed on an electro-servo-hydraulic testing machine under stress ratio (R) of $-1$ (tension-compression mode) and frequency (f) of 10 Hz. The fatigue test was stopped after $1 \times 10^6$ cycles.

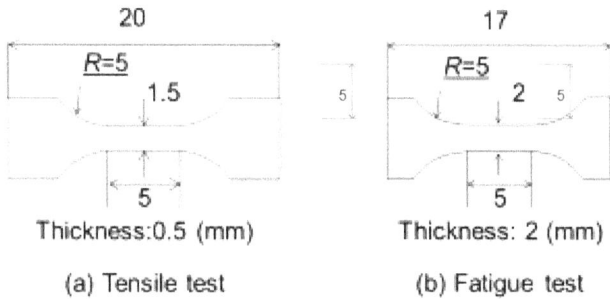

20                                                17
R=5                                               R=5
1.5                                               2
5                                                 5
Thickness:0.5 (mm)                                Thickness: 2 (mm)

(a) Tensile test                                  (b) Fatigue test

*Fig. 4 Schematic drawing of tensile (a) and fatigue (b)*

## 3. Results and discussion

### 3.1 Effect of HIP treatment on microstructure of the EBM-TiAl

Observation results of defects and pores before and after HIP treatment are shown in Figs. 5 and 6.

| as-built | HIP,1100 | HIP,1250°C |
|---|---|---|

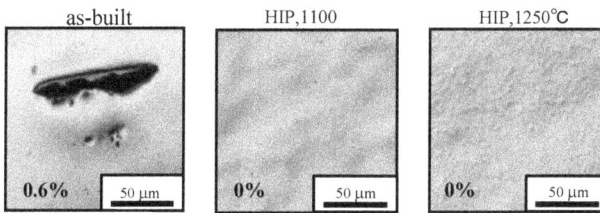

*Fig. 5 Volume fraction of defect change by HIP treatments*

| as-built | HIP,1100°C | HIP,1250°C |
|---|---|---|

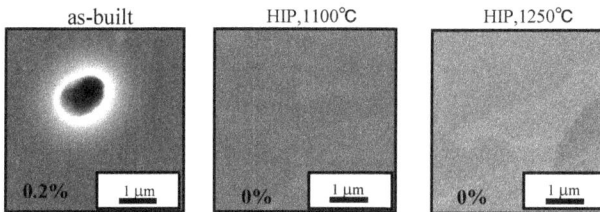

*Fig. 6 Volume fraction of pores change by HIP treatments*

As-build material contains 0.6% of defects and 0.2% of pores. It can be clearly seen that these defects and pores can be drastically eliminated by HIP treatments at 1100 and 1250 °C, indicating the beneficial effect of HIP on the densification of EBM-TiAl. Fig. 7 compares the microstructural change by HIP treatments. $\gamma$ bands were completely deleted by the HIP treatment at 1250 °C, which is conventional HIP temperature for cast TiAl alloys. In this case, volume fraction of $\alpha_2$ phase in duplex-like region increased from 17% to 36%. On the other hand, when HIP treatment was performed at 1100 °C, $\gamma$ band width increased compared with as-built material, and layered microstructure remained. Volume fraction of $\alpha_2$ phase in duplex-like region decreased to 12%.

| as-built | HIP, 1100 °C | HIP, 1250 °C |
|---|---|---|

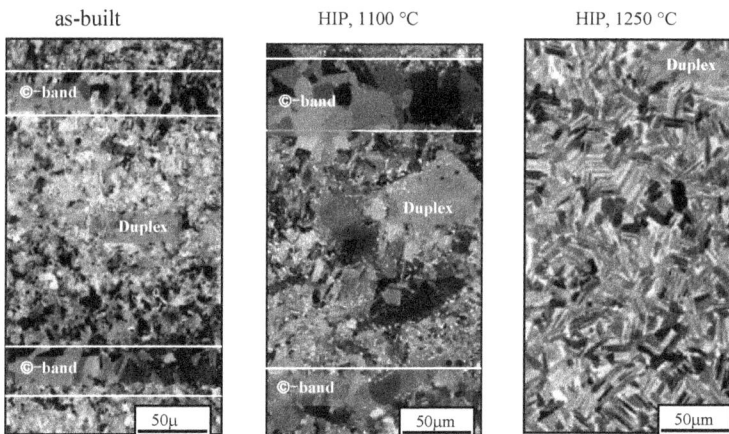

*Fig. 7 Microstructural change in EBM-TiAl by HIP treatment at several*

*Table 2 Change of microstructures*

|  | 1100 °C | 1250 °C |
|---|---|---|
| γ-band | Increased | Decreased strongly |
| α₂ phase in duplex | Decreased | Increased strongly |
| Layered microstructure | No change | All Duplex |

These microstructural changes can be explained by the phase diagram of Ti-Al system [9], which is shown in Fig. **8**. Comparing the HIP treatments at 1100 °C and 1250 °C, it is clear that in case of 1100 °C, volume fraction of γ phase increases and also that of α₂ phase decreases. For this reason, it is obvious that by choosing the optimum HIP temperature, defects and pores can be effectively eliminated while maintaining the layered microstructure of γ bands and duplex-like region.

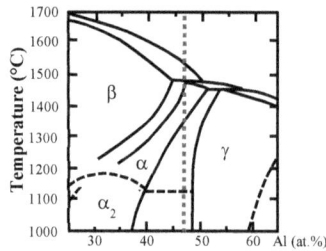

*Fig. 8 Ti-Al binary phase*

### 3.2 Effect of HIP treatment on mechanical properties of the EBM-TiAl

Room temperature tensile properties of as built and HIP treated material are shown in Fig. 9.

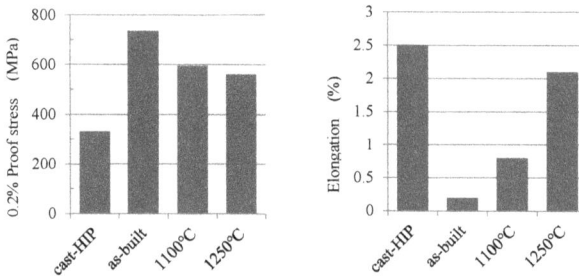

*Fig. 9 RT tensile properties of EBM-TiAl before and after HIP treatment at several temperatures compared to that of cast material after HIP treatment*

HIP treatment at 1100 °C has resulted in the yield strength of 550 MPa or higher, and elongation improved from 0.2% to approximately 1.0%. This improvement in ductility is derived from the increase in γ bands which contributes to the shear deformation. Compared with cast material [10], material HIP treated at 1250 °C shows the equivalent ductility, while maintaining higher yield strength. This high strength is derived from the finer grain size of EBM-TiAl than that of cast material.

Fig. 10 compares the room temperature fatigue properties of as-built and HIP treated material. An improvement is obvious in fatigue life particularly at smaller stress ampli          ly at a high-cycle fatigue life region. It is regarded that drastic reduction of defects and pores by HIP treatment suppresses the initiation of crack nucleus as well as the propagation rate of fatigue cracks. Also material HIP treated at 1100 °C exhibits higher fatigue strength of more than 400 MPa, compared with the one HIP treated at 1250 °C of 300 MPa. This indicates the beneficial presence of γ bands which effectively reduces the crack initiation by promoting strain reduction.

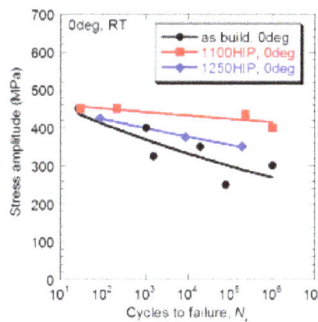

*Fig. 10 Fatigue properties of EBM-TiAl before and after HIP treatment*

From the results of this study, it is obvious that HIP treatment at optimum conditions can improve the mechanical properties of EBM-TiAl significantly. Since the as-build material manufactured in a 45 degree direction exhibits a high elongation of 2.5% or higher, it is expected that further improvements in tensile ductility as well as fatigue properties are possible with the adoption of HIP treatments.

## 4. Conclusions

Cylindrical rods of Ti-48Al-2Cr-2Nb alloy were additively manufactured by EBM method, and microstructural and mechanical properties of HIP treated materials were investigated. The main findings are briefly summarized below.

1. With HIP treatments at 1100 and 1250 °C, defects and pores found in as-build samples can be effectively eliminated.

2. It is possible to maintain the layered microstructure of γ bands and duplex-like region by performing HIP at 1100 °C, where γ band width increases while volume fraction of $\alpha_2$ phase in duplex-like region decreases. On the other hand, HIP treatment at 1250 °C could not maintain the layered microstructure.

3. Mechanical properties of EBM-TiAl can be significantly increased through HIP treatments at 1100 or 1250 °C. Especially, HIP treatment at 1100 °C brings yield strength of more

than 550 MPa and at the same time tensile elongation superior to as built material. Fatigue properties are also improved.

HIP treatment is therefore a key manufacturing step for additively manufactured parts not only for TiAl but for other alloys to achieve high reliability and full density of fine microstructure at the same time.

**References**

[1]  B.P. Bewlay, S. Nag, A. Suzuki, M.J. Weimer, TiAl alloys in commercial aircraft engines, Mater. High Temp. 3409 (2016) 1–11. https://doi.org/10.1080/09603409.2016.1183068

[2]  J. Aguilar, A. Schievenbusch, O, Kättlitz, Investment casting technology for production of TiAl low pressure turbine blades – Process engineering and parameter analysis, Intermetallics 19 (2011) 757–761. https://doi.org/10.1016/j.intermet.2010.11.014

[3]  Shi-Hai Sun, Yuichiro Koizumi, Shingo Kurosu, Yun-Ping Li, Akihiko Chiba, Effect of phase transformation on tensile behavior of Co-Cr-Mo fabricated by electron–beam melting, Journal of Japanese Society of Powder Metallurgy, 61 (2014) 234–242. https://doi.org/10.2497/jjspm.61.234

[4]  Shi-Hai Sun, Yuichiro Koizumi, Shingo Kurosu, Yun-Ping Li, Hiroaki Matsumoto, Akihiko Chiba, Build-direction dependence of microstructure and high-temperature tensile property of Co-Cr-Mo alloy fabricated by electron-beam melting (EBM)**,** Acta Materialia, 64 (2014) 154–168. https://doi.org/10.1016/j.actamat.2013.10.017

[5]  Haize Galarraga, Diana A.Lados, Ryan R.Dehoff, Michael M.Kirka, Peeyush Nandwana, Effects of the microstructure and porosity on properties of Ti-6Al-4V ELI alloy fabricated by electron beam melting (EBM), Additive Manufacturing, 10 (2016) 47-57. https://doi.org/10.1016/j.addma.2016.02.003

[6]  M. Todai, T. Nakano, T. Liu, H. Y. Yasuda, K. Hagihara, K. Cho, M. Ueda and M. Takeyama, Effect of building direction on the microstructure and tensile properties of Ti-48Al-2Cr-2Nb alloy additively manufactured by electron beam melting, Additive Manufacturing, 13 (2017) 61–70. https://doi.org/10.1016/j.addma.2016.11.001

[7]  S. Biamino, A. Penna, U. Ackelid, S. Sabbadini, O. Tassa, P. Fino, M. Pavese, P.Gennaro, C. Badini, Electron beam melting of Ti-48Al-2Cr-2Nb alloy: microstructure and mechanical properties investigation, Intermetallics 19 (2011) 776–781. https://doi.org/10.1016/j.intermet.2010.11.017

[8]  S.Morokoshi, H.Masuo, et al Mechanical properties of Ti-6Al-4V materials prepared by additive manufacturing technology and HIP process, (2014)

[9]  D.M. Dimiduk, D.B. Miracle, Y.W. Kim, M.G. Mendiratta, Recent progress on intermetallic alloys for advanced aerospace systems, ISIJ Int. 31 (1991) 1223–1234. https://doi.org/10.2355/isijinternational.31.1223

[10] Y.-M. Kim, Ordered intermetallic alloys, part III titanium aluminides, JOM, 46 (1994) 30-39. https://doi.org/10.1007/BF03220745

Hot Isostatic Pressing – HIP'17                                    Materials Research Forum LLC
Materials Research Proceedings **10** (2019) 121-127      doi: http://dx.doi.org/10.21741/9781644900031-17

# Overview of Properties, Features and Developments of PM HIP 316L and 316LN

Martin Östlund[1,a] * and Tomas Berglund[2,b]

[1]Sandvik Materials Technology, Storgatan 2, 811 34 Sandviken, Sweden

[2]Sandvik Powder Solutions, Kontorsvägen 1, 735 31 Surahammar, Sweden

[a]martin.ostlund@sandvik.com, [b]tomas.berglund@sandvik.com

**Keywords:** Powder Metallurgy, HIP, 316L, 316LN, Oxides, Inclusions, Microstructure, Impact Toughness, Tensile Properties, PPB

**Abstract.** PM HIP 316L is an alloy that is of increased interest for nuclear applications since its recent ASME code case approval. Over the years, comprehensive data and understanding of the properties and features have been collected and evaluated which will be summarized in this article. Since the early developments of the PM HIP technology it has been observed that PM HIP alloys generally exhibit higher yield strengths compared to their conventional counterparts, a feature that applies well for 316L/LN. In this article this is demonstrated, both by using the Hall-Petch relationship as well as Pickering's and Irvine's empirically derived relationship between composition and grain size for austenitic stainless steels. Furthermore, a mechanism generating the increased yield strength in PM HIP vs conventionally manufactured 316L and 316LN will be proposed. Results also show that low oxygen contents itself is not a guarantee for good or increased performance in form of mechanical properties, but that there are other features that is of similar or perhaps even higher importance in order to achieve good properties. The results of this article include microstructural properties derived from EBSD measurements as well as tensile and impact properties in a wide range of test temperatures of PM HIP 316L and 316LN from several powder batches manufactured at different locations and processed with various HIP and heat treatment procedures. Finally, some results regarding creep properties of PM HIP 316L is presented.

## Introduction

Austenitic stainless steel 316L is one of the most commonly known and used stainless steel grades and the performance and properties of this alloy in different product forms is well known. Powder Metallurgical manufacturing via Gas Atomization and Hot Isostatic Pressing is a manufacturing technology known to generate isotropic microstructures, high cleanliness and often improved mechanical properties. In light of the recent ASME code case approval for PM HIP 316L [1], the properties of this alloy via this manufacturing process has become of increasing interest [2-4]. This article will give an overview of the properties of PM HIP 316L/316LN, how properties can be affected by varying manufacturing process parameters and compare how they differ from the conventionally manufactured counterparts.

## Microstructure

One of the large benefits with PM HIP manufacturing is that the microstructures of the manufactured components are homogeneous, isotropic and have high cleanliness. All these features apply also for PM HIP 316L/316LN and translates into excellent ultrasonic inspectability [4]. Regarding cleanliness, the clear majority of the non-metallic inclusions found in PM HIP 316L/316LN are well below 2.8 μm in size and are predominately constituted by

Hot Isostatic Pressing – HIP'17                                    Materials Research Forum LLC
Materials Research Proceedings 10 (2019) 121-127        doi: http://dx.doi.org/10.21741/9781644900031-17

oxides [2,3]. The oxides can originate either from the melt which are later trapped within the powder particles (bulk oxides) or from the surface oxide layer and oxide particles formed on each powder particle surface after solidification (surface oxides) [3,5]. The latter are often referred to as PPB (Prior Particle Boundary) inclusions in the HIPed microstructure and can form a network that affect ductility and toughness adversely if they are present in large amounts. The general perception from manufacturing experience is that formation of detrimental PPB inclusion networks is not an issue for PM HIP 316L/316LN if properly processed during manufacturing. Inclusions in PM HIP 316L/316LN have been observed to pin grain boundaries and affect grain size [4]. In Fig. 1 a general microstructure displaying grain size and grain orientation derived from EBSD (a) and a SEM image of small inclusions at high magnification can be seen (b) [2, 3]. In Fig. 2 the size distribution of non-metallic inclusions in different batches of PM HIP 316L and 316LN as well as conventionally manufactured 316L (hot rolled Ø50 mm bar) can be observed [2, 3].

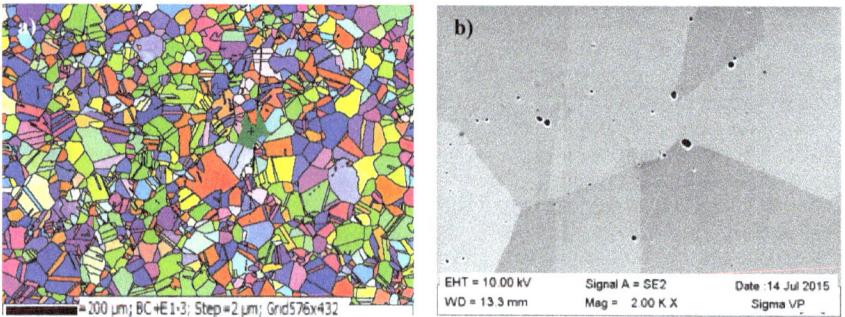

*Fig. 1. EBSD map of grain size and orientation at 100x magnification (a) and small inclusions some pinning grain boundaries, at 2000x magnification (b) in PM HIP 316L [2, 3]*

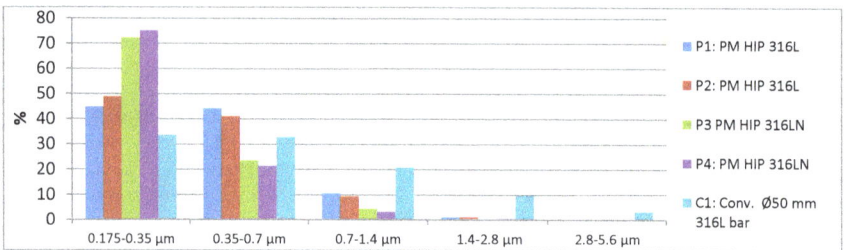

*Fig. 2. Size distribution of inclusions in PM HIP & conventional 316L/316LN [2, 3].*

**Mechanical properties**
Table 1 shows room temperature mechanical properties for PM HIP 316L components with wall thicknesses 60 - 600 mm and weights 10 - 3000 kg. All samples were HIPed at 1150°C/3 h/100 MPa and heat treated at 1050°C for 1.5 - 8 hours followed by water quenching.

**Table 1.** *Average mech. properties ± st. deviation at room temperature for PM HIP 316L.*

| Samples [#] | Rp0.2 [MPa] | UTS [MPa] | A [%] | Impact Toughness [J] |
|---|---|---|---|---|
| 37 | 275 ± 13 | 582 ± 20 | 60 ± 2 | 204 ± 23 |

Looking at the tensile properties of PM HIP 316L and 316LN [2, 3], it appears as if the yield strength is generally higher than for conventionally manufactured counterparts [2, 8-10]. Pickering and Irvine et. al. derived an empirical relationship between composition and microstructural parameters and yield strength for austenitic stainless steels; Eq. 1 [6,7].

$$Rp0.2 \ (MPa) = 15.4[4.4+23(C)+1.3(Si)+0.24(Cr)+0.94(Mo)+1.2(V)+0.29(W)+2.6(Nb)+$$
$$1.7(Ti)+0.82(Al)+32(N)+0.16(delta \ ferrite)+0.46d^{-0.5}]. \tag{1}$$

In Eq. 1 elements are in weight percent, delta ferrite in percent and $d$ is the linear intercept grain size in millimeters. Fig. 3 displays both measured and predicted yield strength at room temperature per Pickering and Irvine et. al. of different PM HIP 316L and 316LN batches as well as for conventionally manufactured counterparts found in literature [8-10]. P4 is a PM HIP 316LN sample characterized identically as P3. As can be observed the Pickering-Irvine prediction is relatively accurate for conventional 316L while the yield strength of the PM HIP samples is consistently underestimated. This is an indication that the PM HIP samples exhibit a strengthening contribution from other factors than composition, delta ferrite and grain size.

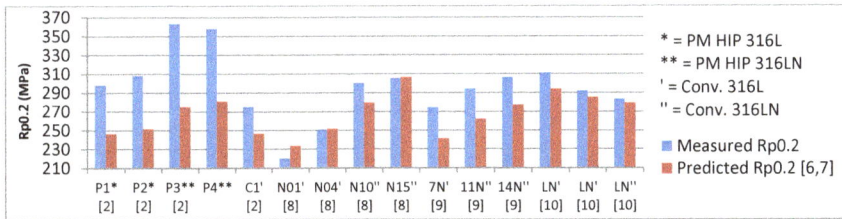

**Fig. 3.** *Measured and predicted yield strength per Pickering and Irvine [6,7] at room temp.*

The main strengthening contributors per to Eq. 1 is grain size, N and C content. If the strengthening contribution from these factors are subtracted from the measured yield strength, a theoretical yield strength (denoted as Rp0.2-n) normalized with regard to these parameters should be obtained. Fig. 4 shows a plot of Rp0.2-n versus amount of oxygen containing inclusions >0.175 μm per $mm^2$. As can be observed there is a good correlation between these parameters, indicating a strengthening effect from the oxygen containing inclusions in PM HIP 316L/316LN which is not present in the conventionally manufactured 316L.

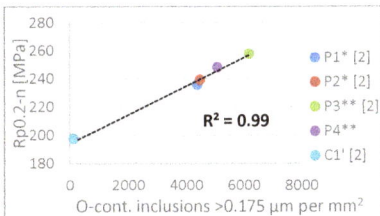

**Fig. 4.** *Rp0.2-n vs. O-cont. incl. per $mm^2$*

In eight different PM HIP 316LN samples of similar composition, the grain size was measured with a HKL Nordlys EBSD and the amount of inclusions >0.175 μm were measured using automated SEM-EDS with a Zeiss Sigma FEG-SEM

Hot Isostatic Pressing – HIP'17                                        Materials Research Forum LLC
Materials Research Proceedings **10** (2019) 121-127          doi: http://dx.doi.org/10.21741/9781644900031-17

using the Aztec and Inca softwares provided by Oxford Instruments. According to the Hall-Petch relationship shown in Eq. 2 there should be a linear correlation between yield strength and grain size [11,12]. However, this correlation was relatively poor for these samples as can be observed in Fig. 5.

$$\sigma_y = \sigma_0 + \frac{k_y}{\sqrt{d}} \tag{2}$$

In Eq. 2 $\sigma_y$ is the yield strength, $\sigma_0$ is the intrinsic yield strength (i.e. the yield strength of the material with infinitely large grain size, also called internal friction stress), $k_y$ is a material specific constant, and d is the mean intercept grain size. The yield strength contribution from the grain size of the samples were calculated in which $k_y$ was chosen to be 164 MPa·$\mu m^{-0.5}$ as derived for 316L [13]. The calculated grain size contribution was then subtracted from the measured yield strength for each sample to obtain the intrinsic yield strength/internal friction stress, $\sigma_0$. In Fig. 6 the intrinsic yield strength/internal friction stress $\sigma_0$ is plotted versus the ECD (Equivalent Circle Diameter) of oxides larger than 0.175 $\mu m$. As can be observed there is a good correlation between these parameters, indicating a strengthening effect from the oxide inclusions which could account for yield strength variations aside from grain size.

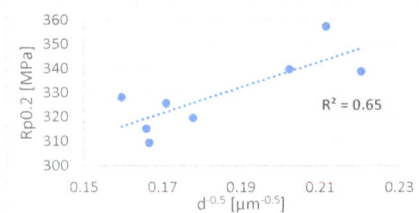

**Fig. 5.** *Yield strength vs. inverted square root of the mean intercept grain size.*   **Fig. 6.** *Intrinsic yield strength $\sigma_0$ vs oxygen containing inclusions >0.175 $\mu m$ per $mm^2$.*

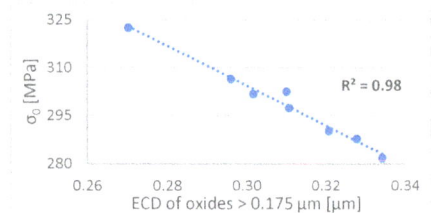

In both cases where PM HIP 316L/316LN samples were compared with conventional counterparts utilizing Irvine and Pickering´s relationship and where eight PM HIP 316LN samples were compared with the Hall-Petch relationship, results indicate a strengthening mechanism by oxygen-containing inclusions in the PM HIP samples. A general feature of PM HIP alloys is that they generally exhibit higher yield strength compared to conventional counterparts, a feature that is valid for PM HIP 316L and 316LN. A possible explanation for this could be that relatively large amounts of small oxygen containing inclusions act as small precipitates impeding dislocation movements during tensile strain, i. e. Orowan strengthening. Such strengthening effect should be stronger for oxygen containing inclusions smaller than 0.175 $\mu m$, but these are more difficult to characterise qualitatively and quantitatively. This theory is strengthened by the observation that the yield strength is reduced by 6-7 % in PM HIP 316LN when the oxygen content is reduced by 47 - 55% [14]. The samples with lower yield strength exhibited larger grain size, but utilizing Eq. 1 and 2. of this study it can be estimated that this can only account for a small amount of the yield strength reduction.

The impact toughness of PM HIP materials is a topic often discussed. Recently there has been raised concern as to why PM HIP 316L seems to drop in impact toughness at cryogenic temperatures [2,3,15]. It appears as if this decrease in impact toughness at cryogenic temperatures is caused by the relatively large amounts of small inclusions. Inclusions also affect impact toughness at room temperature, but due to the strength increase and ductility decrease of the matrix at lower temperatures the inclusions become more detrimental [3]. The impact toughness for several batches of PM HIP 316L/316LN and conventional 316L from different manufacturers at temperatures -196°C – 300°C can be seen in Fig. 7. As can be observed the impact toughness drops at -100°C and -196°C for the PM HIP samples while this is not the case for the forged 316L. Another observation is that the PM HIP materials can meet and exceed the conventional materials at and above room temperature.

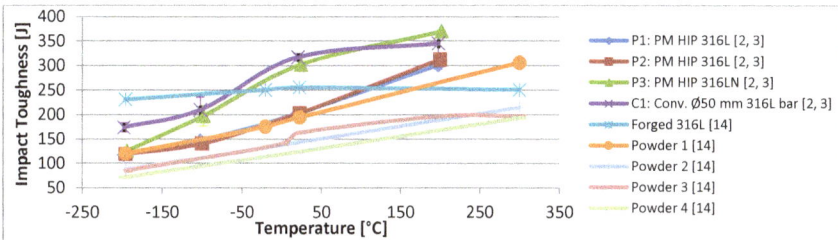

*Fig. 7. Impact toughness for PM HIP and conventional 316L/LN between -196 to 300°C.*

In a similar study for PM HIP 316LN the impact toughness at -196°C increased by ~260 % (from 93 to 243 J), highlighting that oxides are a larger issue at cryogenic temperatures [17]. However, the total oxygen content is not a conclusive indicator on how the materials will perform regarding impact toughness. As explained previously, the total oxygen content in PM HIP materials originate both from bulk oxides and surface oxides [3,5]. The latter source of oxygen has a more detrimental effect on impact toughness as it is known to form a network of oxides on the PPBs if the surface oxygen content is high. PM HIP 316L/LN is known to have a ductile fracture, and voids are normally nucleated around inclusions during deformation [2]. This

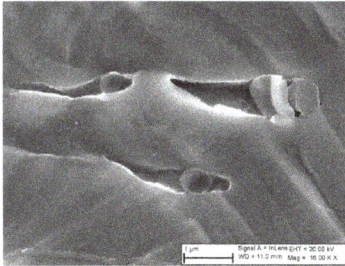

*Fig. 8. In-situ SEM tensile test showing void nucleation and growth around inclusions for PM HIP 316L at 16000x magnification [2].*

has been observed in in-situ SEM tensile testing studies of PM HIP 316L of which an example can be seen in Fig. 8. These voids grow during further deformation and ultimately coalesce with each other leading to fracture [2,3,18]. Having larger amounts of oxides in the microstructure as an effect of higher oxygen content will result in increased number of sites for void nucleation and reduced space for voids to grow without coalescing with adjacent voids, thus accelerating the fracture propagation. In the case of PPB oxides, void coalescence occurs almost immediately after void nucleation due to the vicinity of each PPB oxide leading to significantly reduced impact toughness.

Fig. 9 shows an example of how the impact toughness can vary between different samples of PM HIP 316LN even though oxygen contents are similar (a), and how the impact toughness can vary depending on manufacturing process parameters for the same batch (b). Relatively large differences in impact toughness can be observed for PM HIP 316LN between different batches and process parameters which indicate that the total oxygen content is not the only parameter to indicate this property in PM HIP materials. Note that some samples reach close to 400 J which is similar to the previously mentioned PM HIP 316L with greatly reduced oxygen content (22 ppm) [16]. This highlights that good impact toughness can be achieved with 100 ppm oxygen content, i.e. without having to greatly reduce oxygen content.

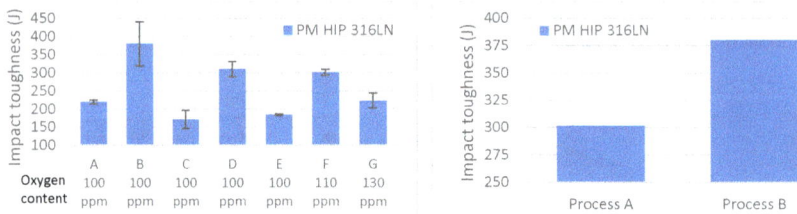

***Fig. 9.*** *Samples with similar oxygen content (a), effect of process parameters on same batch.*

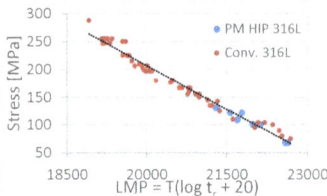

***Fig. 10.*** *Creep properties.*

In Fig. 10. creep results for a PM 316L sample of the same batch as P1 [2, 3] HIPed at 1125°C/100 MPa for 2 hours is presented as stress versus Larsson-Miller parameter (LMP) and compared to data for conventionally manufactured 316L. T is test temperature in Kelvin and $t_r$ is hours to rupture. As can be observed the results are similar for PM HIP and conventional 316L. Test specimens were connected in series in test cells which were loaded prior to heating. The samples were at different instances removed from the furnace, unloaded and cooled down for measurements. No continuous measurements of load and elongation was available in the test setup.

## Summary

PM HIP 316L/LN exhibits a homogeneous and isotropic microstructure with high cleanliness. Non-metallic inclusions found in the microstructure are small and relatively evenly distributed. Oxygen containing inclusions can seemingly affect the mechanical properties, both positively as in the case of yield strength, and adversely in some cases for impact toughness. The total oxygen content in PM HIP 316L/LN can on a broader scale indicate impact toughness levels, but results of this study shows that it is not the only parameter for this. Results presented in this study shows that excellent properties can be achieved for PM HIP 316L/LN at moderate oxygen levels if processed correctly.

## References

[1] ASME Boiler and Pressure Vessel Code Case N-834

[2] T. Berglund, M. Östlund, L. Larsson, Impact toughness of PM HIPed vs. conventional 316L, Proceedings of the Euro PM2015 Congress (2015).

[3] T. Berglund, M. Östlund, Impact toughness for PM HIP 316L at cryogenic temperatures, ASME PVP Conference 2016, Vol 6A: Materials and Fabrication (2016).

[4] D. Gandy, Program on Technology Innovation: Manufacture of Large Nuclear and Fossil Components Using Powder Metallurgy and Hot Isostatic Processing Technologies, Technical Report 1025491, Electric Power Research Institute (2012).

[5] L. Nyborg, Ytsyre och dess inverkan på mekaniska egenskaper hos hetpressade PM-material, Jernkontorets Forskning, Nr. 681, Serie D, TO 80-26, 8064/88, (1993).

[6] F. B. Pickering, Physical metallurgy of stainless steel developments, International Metals Reviews. 21 (1976) 227-277.

[7] K. J. Irvine, T. Gladman, F. B. Pickering. Strenght of austenitic stainless steels, Journal of the Iron and Steel Institute 207 (1969) 1017-1028.

[8] D. W. Kim, W-S Ryu, J. H. Hong, S-K Choi, Effect of nitrogen on the dynamic strain ageing behaviour of type 316L stainless steel, Journal of Materials Science 33 (1998) 675-679. https://doi.org/10.1023/A:1004381510474

[9] J. G. Kumar, M. Chowdary, V. Ganesan, R.K. Paretkar, K. Bhanu Sankara Rao, M.D. Mathew, Nuclear Engineering and Design 240 (2010) 1363-1370. https://doi.org/10.1016/j.nucengdes.2010.02.038

[10] M. Nyström, U. Lindstedt, B. Karlsson, J-O. Nilsson, Influence of nitrogen and grain size on deformation behaviour of austenitic stainless steel, Mater. Sci. Technol. 13 1997 560-567. https://doi.org/10.1179/mst.1997.13.7.560

[11] E. Hall, The deformation and ageing of mild steel: III. Discussion of results, Proceedings of the Physical Society, Section B, 64 (1951) 747. https://doi.org/10.1088/0370-1301/64/9/303

[12] N. Petch, The cleavage strength of polycrystals, J Iron Steel Inst. 174 (1953) 25.

[13] N. Hirota, F. Yin, T. Azuma, T. Inoue, Yield stress of duplex stainless steel specimens estimated using a compound Hall-Petch equation, Sci. Technol. Adv. Mat. 11 (2010) 025004. https://doi.org/10.1088/1468-6996/11/2/025004

[14] A. Lind, J. Sundström, A. Peacock, The effect of reduced oxygen content powder on the impact toughness of 316L steel powder joined to 316 steel by low temperature HIP, Fusion Engineering and Design 75–79 (2005) 979–983. https://doi.org/10.1016/j.fusengdes.2005.06.301

[15] D. Cédat et. al. Advanced mechanical properties of P/M 316L stainless steel material, Presentation at Euro PM2012, September 16-19 (2012) Basel, Switzerland.

[16] A. Angré. A. Strondl, The effect of drastically lowered oxygen levels on impact strength for HIP'ed 316L material, Proc. Of the 11th Int. Conf. on HIP (2014) 162-168.

[17] J. Sundström, Influence of oxygen content on impact strength of hot isostatically pressed PM superduplex stainless steel and 316L stainless steel, Swerea Kimab IM-2005-509 (2005).

[18] W.T. Becker, R.J. Shipley. Volume 11: Failure Analysis and Prevention, ASM Handbook, ASM International 2002 587-626.

Materials Research Forum LLC
doi: http://dx.doi.org/10.21741/9781644900031-18

# Effect of Processing Parameters on Intermetallic Phase Content and Impact Toughness for Super Duplex Alloy PM HIP Sandvik SAF 2507™

Martin Östlund[1,a] *, Linn Larsson[1,b] and Tomas Berglund[2,c]

[1]Sandvik Materials Technology, Storgatan 2, 811 34 Sandviken, Sweden

[2]Sandvik Powder Solutions, Kontorsvägen 1, 735 31 Surahammar, Sweden

[a]martin.ostlund@sandvik.com, [b]linn.larsson@sandvik.com, [c]tomas.berglund@sandvik.com

**Keywords:** Powder Metallurgy, HIP, Super Duplex Stainless Steel, Intermetallic phases, Sigma Phase, Impact Toughness

**Abstract.** PM HIP is a widely applied manufacturing technology to produce thick walled and complex shaped duplex and super duplex stainless steel (DSS and SDSS) components for the petrochemical as well as the oil and gas industry. The PM HIP process offers the advantage of a fine-grained microstructure which generates an increased resistance to HISC (Hydrogen Induced Stress Cracking) as well as higher yield strength. A limiting factor when producing thick walled components of DSS and SDSS alloys is the precipitation of brittle intermetallic phases which results in decreased corrosion resistance and impact toughness if high enough fractions are precipitated. The precipitation of intermetallic phases is a diffusion controlled process that may take place during quenching following solution annealing if the cooling rate is too slow. The cooling rate during quenching is mainly depending on the section thickness of the component, where large sections are subjected to slower cooling and thus increased intermetallic phase precipitation. In this article, it is shown that a coarser PM HIP microstructure results in lower contents of intermetallic phases after water quenching. However, despite of the lower intermetallic phase content the impact toughness is not improved and this is explained by the fracture mechanisms as shown by instrumented impact testing and fracture surface analysis.

## Introduction

Sandvik SAF 2507™ is a super duplex stainless steel characterized by excellent resistance to stress corrosion cracking, pitting and crevice corrosion, general corrosion and high mechanical strength. Increasing water depths (increasing pressures) and increasing process temperatures in PM HIP applications for the oil and gas sector results in designs with increasingly large wall thickness. A limiting factor when it comes to increased wall thickness is the formation of brittle intermetallic phase during water quenching following heat treatment. The intermetallic phases nucleate and grow primarily in ferrite grain boundaries and ferrite/austenite phase boundaries in the approximate temperature interval 600 - 1000°C [1,2]. The thicker the manufactured component is, the longer times the center part of each section is subjected to the temperature interval in which intermetallic phase is formed during water quenching. Even smaller amounts of intermetallic phase content may affect the impact toughness adversely for DSS and SDSS components. This study was conducted to investigate if a coarser microstructure (i. e. reduced grain and phase boundary area) obtained by higher HIP temperature could result in lower amounts of intermetallic phase along with improved impact toughness of thick walled components of PM HIP SAF 2507.

## Experimental

Two mild steel capsules with dimensions Ø133x250 mm and two with dimensions Ø236x250 mm were filled with SAF 2507 powder with composition per Table 1. The filled capsules were evacuated after which one of each capsule type were HIPed at 1150°C and 100 MPa with 3 hours holding time. The remaining two capsules were HIPed in another HIP cycle at 1250°C and 100 MPa with 3 hours holding time. The HIPed capsules were tested with regard to Argon content to verify that no capsule leakages occurred during HIP. Once it was verified that no detectable argon was present in the HIPed capsules they were heat treated. Capsule ID, HIP and heat treatment details can be seen in Table 2.

*Table 1. Chemical composition [wt%] of SAF 2507 powder batch.*

| C | Si | Mn | P | S | Cr | Ni | Mo | Cu | N |
|---|----|----|---|---|----|----|----|----|---|
| 0.015 | 0.41 | 0.80 | 0.018 | 0.002 | 25.0 | 6.85 | 3.82 | 0.09 | 0.28 |

*Table 2. Capsule, HIP and heat treatment details.*

| Capsule ID | HIPed capsule size | HIP parameters | Heat treatment parameters |
|------------|--------------------|-----------------|----------------------------|
| 1191-1 | ~Ø120x220 mm | 1150°C/3h/100 MPa | 1070°C/4h/WQ |
| 1192-1 | ~Ø210x220 mm | 1150°C/3h/100 MPa | 1070°C/5.75h/WQ |
| 1191-2 | ~Ø120x220 mm | 1250°C/3h/100 MPa | 1070°C/4h/WQ |
| 1192-2 | ~Ø210x220 mm | 1250°C/3h/100 MPa | 1070°C/5.75h/WQ |

Three Charpy V-notch impact test bars were prepared from each of the HIPed and heat treated capsules at half height and surface, half radius and center position respectively. The manufactured test bars were tested by instrumented impact testing at -46°C per ASTM 2298. The CPT (Critical Pitting Corrosion Temperature) was measured per ASTM G150 on tested impact test bars from surface and center locations of capsules 1192-1 and 1192-2.

EBSD data collection was performed at 500x magnification on the non-deformed microstructures of ruptured impact test bars at surface, half radius and center position of the Ø210 mm capsules (1192-1 & 1192-2). The amount of austenite, ferrite and sigma phase was measured. The grain size was determined as area weighted average equivalent circle diameter (ECD) and as linear intercept grain size was from the EBSD images using 50 equidistant horizontal and vertical lines. Grain detection was performed disregarding sigma 3 twin boundaries in both cases. The EBSD data collection details can be seen in Table 3 and Fig. 1.

*Table 3. EBSD data collection details.*

| Parameter | Setting |
|-----------|---------|
| Camera resolution | 461 x 345 pixels |
| Binning | 4 x 4 (160x120 pixels) |
| Exposure time | 13.8 ms |
| Gain | 15 |
| Band detection | 12 |
| Hough resolution | 60 |
| Step size | 0.5 µm |
| Image size | 0.23 x 0.17 mm = 0.039 mm$^2$ |

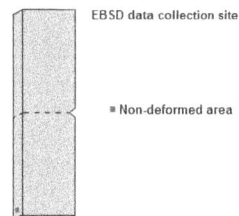

*Fig. 1. EBSD site.*

Hot Isostatic Pressing – HIP'17                                    Materials Research Forum LLC
Materials Research Proceedings **10** (2019) 128-134          doi: http://dx.doi.org/10.21741/9781644900031-18

In addition to EBSD quantification of sigma phase and grain size measurements, the amount of intermetallic phase and austenite spacing was determined on the center samples of capsules 1192-1 and 1192-2 using image analysis. The samples were polished and etched in Murakami´s etchant after which image analysis at 500x magnification was conducted on 20 random fields of view.

## Results

The results from the impact testing at -46°C showed similar impact toughness at centre, half radius and surface positions for the investigated capsule sizes regardless of HIP-temperatures. Although there were no large differences, slightly lower impact toughness is indicated for capsules HIPed at 1250°C compared to the capsules HIPed at 1150°C. Previous studies conclude that the reduction in impact toughness at centre position of the Ø120 mm capsules and half radius and centre position of the Ø210 mm is due to intermetallic phase content [1,3,4]. The results from the impact testing can be seen in Fig. 2 along with previous results for the same powder batch [3] where the average values of three samples are presented with the standard deviation as error bars.

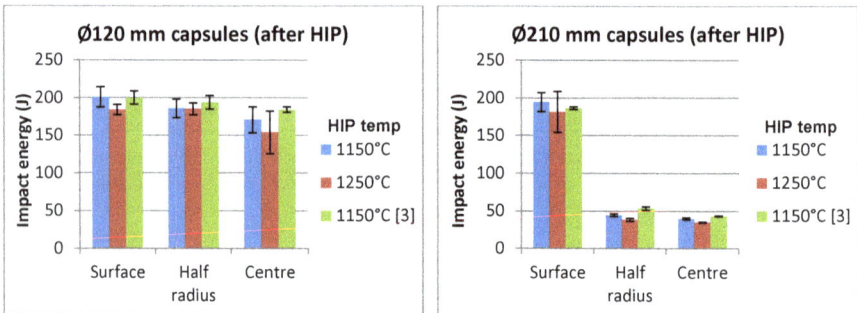

*Fig. 2. Impact toughness results at -46°C for Ø120 mm and Ø210 mm capsules HIPed at 1150°C and 1250°C.*

Some differences can be observed in the load vs. deflection curves of selected test specimens of each capsule obtained from the instrumented impact testing which is shown in Fig. 3. For surface samples in both Ø120 mm and Ø210 mm capsules it can be observed that HIP at 1150°C generates fully ductile fracture while HIP at 1250°C generates a ductile-brittle fracture. For the mentioned samples HIPed at 1250°C a brittle region can be observed (vertical drop) after crack initiation (i. e. after peak load) leading to lower crack propagation energy and lower impact toughness in total. For half radius and center samples in the Ø210 capsule, it can be observed that all samples exhibit a ductile-brittle fracture. Fracture surfaces of impact test bars tested at -46°C from surface position in 180x70x50 mm capsules HIPed at 1150°C and 1250°C followed by 1070°C/2h/WQ heat treatment is shown in Fig.4 and Fig. 5 respectively. These samples were produced from the same powder batch as the samples of this study [5]. In these figures, it can be observed that specimens HIPed at 1150°C exhibits a fully ductile dimple fracture while samples HIPed at 1250°C exhibits ductile dimple fracture with local areas of brittle fracture appearing to be fractures along grain and/or phase boundaries or alternatively cleavage fracture.

The amount of intermetallic phase and austenite spacing measured by image analysis at the center position of the Ø210 mm capsules can be seen in Table 4. All values are expressed as average values ± standard deviation. As can be seen it appears as if the capsule HIPed at 1250°C contains lower amounts of intermetallic phase, although the standard deviations are overlapping.

***Table 4.*** *Intermetallic phase content (area percent) and austenite spacing.*

| Capsule ID | HIP temperature | Intermetallic phase [%] | Austenite spacing [μm] |
|------------|-----------------|--------------------------|-------------------------|
| 1192-1 | 1150°C | 0.197 ± 0.096 | 8.7 ± 5.2 |
| 1192-2 | 1250°C | 0.112 ± 0.076 | 13.2 ± 8.1 |

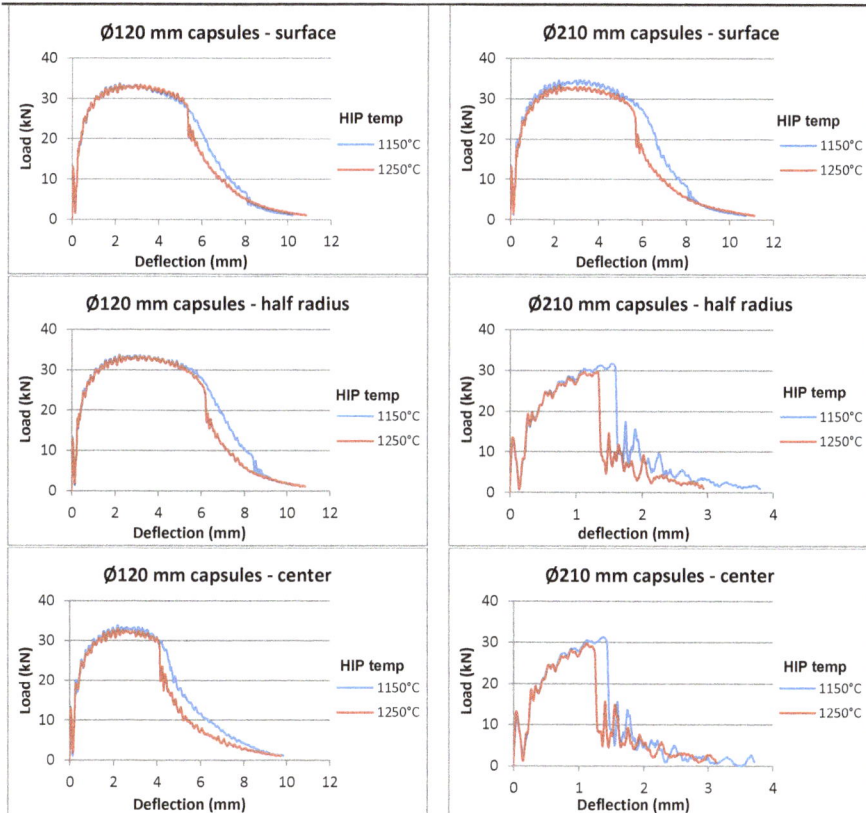

***Figure 3.*** *Load vs. deflection curves from the instrumented impact testing at -46°C.*

***Fig.4.*** *Fracture surface of impact specimen HIPed at 1150°C, 200x magnification [5]*

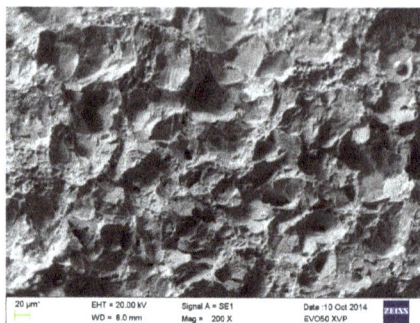

***Fig. 5.*** *Fracture surface of impact specimen HIPed at 1250°C, 200x magnification [5]*

Grain size measurements from the EBSD data collection at center, half radius and surface location of the Ø210 mm capsules can be seen in Table 5. As can be noted the grain size of the capsule HIPed at 1250°C is approximately 60 – 65 % larger than for the capsule HIPed at 1150°C. The grain size is generally larger in the austenite (FCC) compared to the ferrite (BCC) regardless of HIP temperature. Another observation is that the grain size at the center seems to be slightly smaller (3 – 6 %) than the grain size at half radius and surface positions.

***Table 5.*** *Grain size from the EBSD data collection presented as average values.*

| Capsule ID | Phases | Area weighted ECD [µm] | | | Line intercept [µm] | | |
|---|---|---|---|---|---|---|---|
| | | Surface | Half radius | Center | Surface | Half radius | Center |
| | All | 12.15 | 12.22 | 11.85 | 7.20 | 7.14 | 6.75 |
| 1192-1 | FCC | 12.54 | 12.59 | 12.33 | 7.82 | 7.81 | 7.33 |
| | BCC | 11.66 | 11.69 | 11.18 | 6.54 | 6.39 | 6.14 |
| | All | 20.08 | 19.72 | 19.18 | 11.70 | 11.89 | 11.10 |
| 1192-2 | FCC | 20.91 | 19.72 | 19.99 | 13.45 | 12.62 | 11.44 |
| | BCC | 18.97 | 19.72 | 18.04 | 9.96 | 11.13 | 9.87 |

Images from the EBSD data collection of mentioned samples can be seen in Fig. 6 where the austenite (FCC) phase is marked in blue, ferrite (BCC) in red and sigma phase in black color. From the EBSD data collection results it appears as if the material HIPed at 1250°C generally contains lower amounts of sigma phase. Since the measurements are only conducted on single fields of view there is however no statistical basis to support this. The average CPT from the ASTM G150 corrosion testing is detailed in Table 6. Similar values are obtained for all samples. The results are in line with previous results from both PM HIP SAF 2507 and conventionally produced SAF 2507 (80-90°C) [1,2].

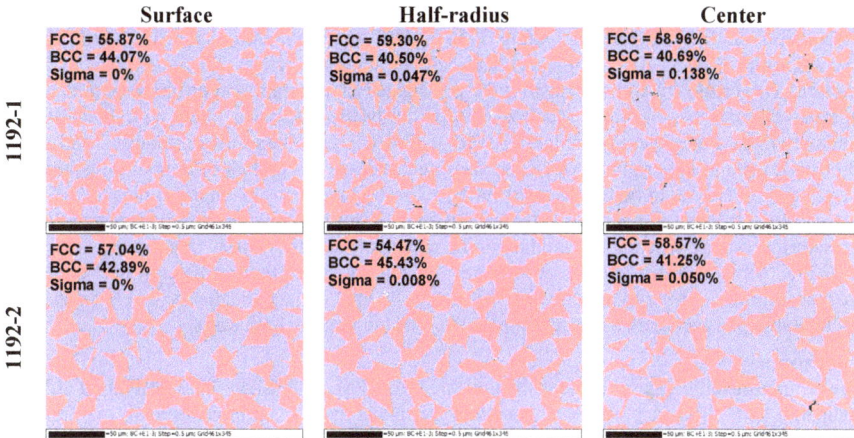

**Fig. 6.** *EBSD maps at center, half radius and surface positions in the Ø210 mm capsules (FCC = blue, BCC = red and Sigma phase = black), 500x magnification.*

**Table 6.** *Average CPT from ASTM G150 testing.*

| Sample ID | Critical Pitting Corrosion Temperature [°C] | |
|-----------|------------------|----------------|
|           | Capsule surface  | Capsule center |
| 1192-1    | 88.0             | 90.5           |
| 1192-2    | 88.4             | 89.2           |

**Discussion**

The results of this study indicate that a lower amount of intermetallic phase is formed in PM HIP SAF 2507 when produced with higher HIP temperature. The measured differences are however small and the standard deviations of the mean values are overlapping. It should be observed that lower amounts of sigma phase are measured with EBSD compared to intermetallic phase content measured by image analysis. A likely explanation to this could be that the EBSD analysis only measures sigma phase while the image analysis measures all intermetallic phases, i.e. also includes any eventual χ-phase. The intermetallic phase content can also be over-estimated due to etching effects while it can be underestimated with EBSD due to the inadequate resolution from the selected step size. There are some non-indexed points in the EBSD maps surrounding the sigma phase areas which suggest this. Beside these factors there is also the obvious difference in number of fields of view (1 vs 20), i.e. the EBSD measurements are more uncertain. A likely explanation for smaller amounts of intermetallic phase in the samples HIPed at 1250°C would be the coarser microstructure. A coarser microstructure leads to reduced grain and phase boundary area and ultimately result in fewer possible locations/smaller area for nucleation and growth of intermetallic phase.

Even though lower amounts of intermetallic phase are indicated for the capsules HIPed at 1250°C, the impact toughness is not improved compared to the capsules HIPed at 1150°C. By observing the load vs. deflection curves from samples free of intermetallic phase (surface samples of Ø120 and Ø210 mm capsules), it can be observed that 1150°C HIP results in a fully ductile fracture while 1250°C HIP results in ductile-brittle fracture. This also coincides with observations at fracture surfaces where 1150°C HIP samples exhibits a fully ductile dimple fracture while 1250°C HIP samples exhibits ductile dimple fracture with local areas of brittle fracture. The brittle fractures seem to have propagated along grain and/or phase boundaries or alternatively they are cleavage fractures. The partial brittle fracture is likely an explanation for the lower impact toughness of samples HIPed at 1250°C, regardless of intermetallics, and seems to be correlated to larger grain size. The exact mechanism causing the brittle fracture is not fully understood and needs to be investigated further. The half radius and center samples of the Ø210 mm capsules is likely to contain the largest amounts of intermetallic phase due to slower cooling rates during water quenching. Even though the results of this study imply that the capsule HIPed at 1250°C contains lower amount of intermetallic phase content, the results in impact toughness are very similar. The explanation for the similarly low impact toughness levels is that the intermetallic phase content of these samples likely is too large in both cases.

**Conclusion**

Lower intermetallic phase content is indicated for PM HIP SAF 2507 HIPed at 1250°C compared to 1150°C. A probable explanation for this is the coarser microstructure, i.e. reduced grain and phase boundary area, which results in fewer locations for nucleation and growth of intermetallic phase. Samples HIPed at 1250°C free of intermetallic phase exhibits a ductile-brittle fracture during impact testing, which manifests itself as partial brittle fracture. This results in lower crack propagation energy and consequently lower impact energy compared to samples HIPed at 1150°C which exhibit fully ductile fracture with corresponding ductile dimple type fracture surface. Even though higher HIP temperatures might reduce the susceptibility towards formation of intermetallic phases, the impact toughness is not necessarily improved. The coarse microstructure itself seem to generate lower impact toughness regardless of intermetallic phase content.

**References**

[1] L. Larsson, M. Bjurström, Properties of PM HIPed SAF 2507 super duplex stainless steel, Proceedings of Euro PM2012 Conference (2012).

[2] J. O. Nilsson, Overview Super duplex stainless steels, Mat. Sc. Tech. 8 (1992) 685-699. https://doi.org/10.1179/mst.1992.8.8.685

[3] L. Larsson, Internal Sandvik R&D technical report, 120894TEA (2012).

[4] L. Larsson, Internal Sandvik R&D technical report, 121625TEA (2012).

[5] M. Östlund, Internal Sandvik R&D technical report, 170270TEA (2017).

Hot Isostatic Pressing – HIP'17            Materials Research Forum LLC
Materials Research Proceedings **10** (2019) 135-141     doi: http://dx.doi.org/10.21741/9781644900031-19

# Oxygen Content in PM HIP 625 and its Effect on Toughness

Tomas Berglund[1,a*], Fredrik Meurling[2,b]

[1]Sandvik Powder Solutions, PO Box 54, 79521 Surahammar, Sweden

[2]Sandvik Materials Technology, TWRP, 81181 Sandviken, Sweden

[a]tomas.berglund@sandvik.com, [b]fredrik.meurling@sandvik.com

**Keywords:** HIP, Wear, Tribology, Metal Matrix Composites, MMC, Wear Resistant

**Abstract.** Oxygen control during powder manufacturing and handling is crucial when manufacturing HIPed parts. The influence of elevated oxygen content on mechanical properties is something that has been debated and investigated for many years. The general consensus in the industry is that oxygen has a very detrimental effect on the toughness of the material if present in excessive amounts.

The detrimental effect of oxygen content on the impact toughness of the material has resulted in HIPed specifications, both existing and under development, with limits on the oxygen content in the material. Many specify a relatively low limit on oxygen content at e.g. 120 ppm which can have adverse effects on yield in powder manufacturing which might increase costs without accomplishing the desired effect of ensuring sufficient toughness. As this study show, oxygen content and chemistry alone is not enough to describe the effect of oxygen content on the HIPed material. Setting a limit at e.g. 120 ppm will not guarantee that one gets better properties or even reaches the desired properties of the material. The study show it is important where the oxygen is located in the powder and to separate bulk oxygen content and the surface oxygen content, where the latter has a more pronounced effect on toughness. In the study four batches of alloy 625 have been investigated, all with only relatively small variations in oxygen content but with drastically different toughness and differences in how oxygen is distributed in the material.

## Introduction

Powder Metallurgical (PM) materials are sensitive to oxygen due to the large surface area of the fine powder. In some PM processes e.g. press & sinter and Metal Injection Molding, oxygen content can be reduced in sintering by performing it in hydrogen. However, when consolidating the material using Hot Isostatic Pressing (HIP) the consolidation occurs with vacuum the capsule which has little or no effect on the oxygen content. Therefore, oxygen control throughout the manufacturing process is important as any adsorbed oxygen cannot be removed in the later stages of manufacturing. Other studies have investigated the influence of oxygen on mechanical properties on HIPed austenitic and duplex stainless steel. In general the studies show a correlation between oxygen content and impact toughness, especially at lower temperatures [1-6]. Usually it is toughness that is reduced by excessive oxygen in the material but also welding properties of the material can be affected.

Currently there are few material specifications on HIPed material and most that exist are project or product specific. There are a few specifications and standards covering PM HIP material e.g. ASTM (A988, A989 and B834), ASME code cases (N-834 and 2840) as well a mention in API 6A. However, more specs are in the works and many of them specify maximum oxygen content in the material. There is a trend to set lower and lower maximum allowable oxygen content which in turn can have a negative effect on price of the produced parts. When

Hot Isostatic Pressing – HIP'17                                    Materials Research Forum LLC
Materials Research Proceedings **10** (2019) 135-141        doi: http://dx.doi.org/10.21741/9781644900031-19

levels are below 120 ppm it gets much more difficult for powder and part manufacturers to meet this and it might not have the desired effect on mechanical properties.

Other studies have shown that properties at levels of oxygen from 120 ppm and below is not necessarily connected to the amount of oxygen, in fact a material with higher oxygen content can have better toughness than a material having significantly lower oxygen content [7, 8]. In this study 4 different Alloy 625 materials, manufactured with Ar or N gas atomizing have been investigated with regards to microstructure and mechanical properties.

## Experimental

**Sample manufacturing**. Manufacturing of the powders was done using gas atomization. Process and powder handling parameters was varied to achieve different distribution of the oxygen in the powder. The atomized powders were sieved at -250 μm prior to filling of the capsules. N and Ar atomized powder are hereafter labeled N625 and A625 respectively.

The powders were filled in rectangular-shaped mild-steel capsules of outer dimensions 180x70x50mm and sheet thickness 2mm, evacuated and sealed and subsequently HIPed in a standard HIP cycle with a plateau at 1150°C temperature, 100 MPa pressure and 3 hours. Testing on all materials was performed in the as-HIPed condition.

**Chemical analysis.** All materials were analyzed with regards to chemical composition in the as-HIPed condition. Ar-testing was done on the capsule filling pipe that was filled with 253MA material. The same procedure that is often used in the industry.

**Mechanical testing.** Tensile testing was performed using ISO 6892-1:2009. Charpy impact toughness testing was performed per ASTM A370-17 at -46°C. Average of three tests is presented.

**Microstructural characterization.** Was performed using Light Optical Microscopy (LOM) as well Scanning Electron Microscopy (SEM) and Energy Dispersive X-ray Spectroscopy (EDS).

## Results

**Chemical analysis.** The chemical analysis for the material in the as-HIPed condition can be seen in table 1. All the material are similar but do contain some minor differences, especially when comparing N and Ar atomized powders. As expected the N-content in the N-atomized powder is significantly higher compared to the Ar-atomized powder. The later does contain higher amounts of the strong nitride forming elements Ti and Al as well as a lower amount of Fe.

*Table 1. Composition of materials in the as-HIPed condition (wt.%).*

|        | Ni   | Cr    | Mo   | Nb   | Fe   | Ti    | Al   | C     | Si   | N     | S     | P      | O      |
|--------|------|-------|------|------|------|-------|------|-------|------|-------|-------|--------|--------|
| A625:1 | Bal. | 21.14 | 8.99 | 3.55 | 1    | 0.22  | 0.21 | 0.007 | 0.02 | 0.007 | 0.002 | 0.003  | 0.0095 |
| A625:2 | Bal. | 21.43 | 9.07 | 3.71 | 1.11 | 0.29  | 0.27 | 0.018 | 0.06 | 0.006 | 0.001 | 0.003  | 0.0128 |
| N625:1 | Bal. | 21.59 | 9.25 | 3.73 | 2.55 | 0.01  | 0.07 | 0.021 | 0.02 | 0.066 | 0.001 | <0.003 | 0.0105 |
| N625:2 | Bal. | 21.74 | 9.16 | 3.69 | 2.38 | <0.02 | 0.03 | 0.015 | 0.02 | 0.1   | 0.001 | 0.003  | 0.0096 |

**Microstructure.** Figure 1 show LOM and backscattered SEM micrographs of the Ar-atomized materials. In the A625:1 material several clusters oxide particles are observed (white spots in figure a, black spots in figure c). Many of these are correlated to the surface of the prior powder particles as they form a semi-continuous network that clearly highlight the spherical shape of the prior powder particles. These so called Prior Particle Boundary particles (PPBs) are considerably smaller than the otherwise occurring particles that are located inside the prior powder particles. In the A625:2 material the semi continuous PPB structure is much less pronounced and the particles generally smaller. Studying the material in SEM and EDS (see figure 1) it is found that for the A625:1 material the oxide particles in the larger clusters appear to be purely alumina without traces of Ti-rich precipitates. The PPBs are decorated with alumina particles but also titanium nitrides and/or oxides as well as combinations of all these three. The occurrence of other precipitates is very limited. Some single particles rich in Mo, Cr, Nb and C are however found that might be some low-carbon containing carbide or possible an intermetallic phase. The size of these latter particles is below 1 μm. In the A625:2 material the PPBs are decorated with alumina particles, sometimes also containing titanium. Inside the old powder particles there are small precipitates enriched in Nb, C, Ti and N, most often situated along grain boundaries and having a thickness below 1 μm.

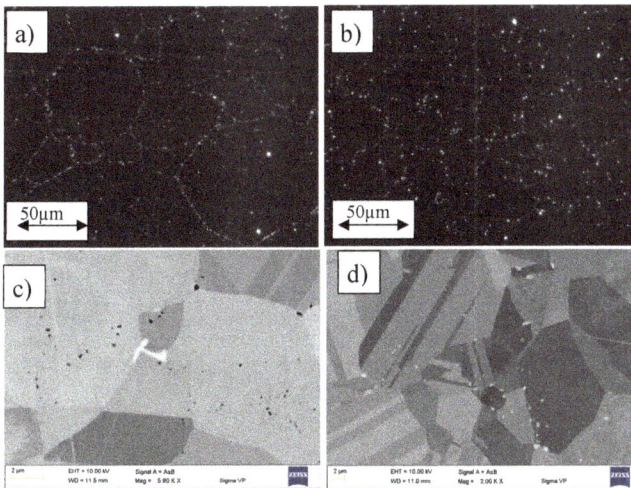

*Figure 1. LOM (a & b) and SEM (c & d) micrographs of A625:1 (a & c) and A625:2 (b & d).*

*Table 2. Mechanical properties.*

|        | $Rp_{0,2}$ [MPa] | $R_m$ [MPa] | A [%] | Z [%] | CVN [J] |
|--------|------|------|------|------|------|
| A625:1 | 463  | 909  | 41   | 35   | 71   |
| A625:2 | 500  | 948  | 49   | 48   | 93   |
| N625:1 | 464  | 946  | 37   | 31   | 59   |
| N625:2 | 477  | 966  | 43   | 43   | 87   |

Studying the microstructure of N625:1 in LOM, PPB structures can clearly be seen (figure 2a). It is also clear that the inside of the powder particles contains a lower amount of precipitates. The N625:2 material (figure 2b) in comparison is the opposite, the PPB structure is not as apparent and the inside of the particles contain a significantly higher number of oxides. In SEM/EDS (figure 2 c&d) the particles decorating the PPBs can be identified as Al-rich oxides. The precipitates inside the powder particles are rich in N, Nb, Cr and Mo. They are elongated and a few μm in size.

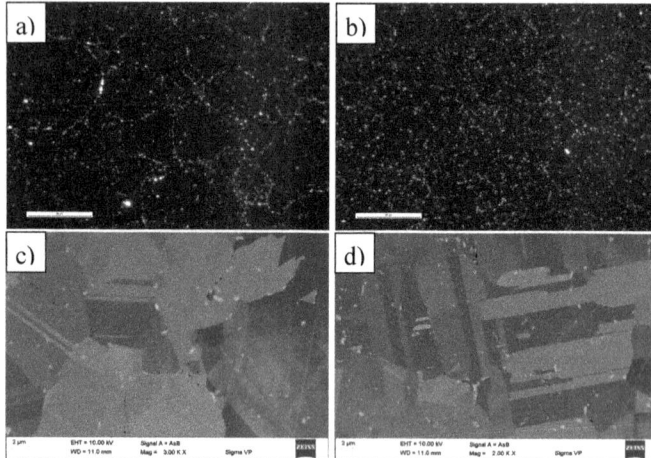

*Figure 2. LOM (a & b) and SEM (c & d) micrographs of N625:1 (a & c) and N625:2 (b & d).*

**Mechanical properties**

The results from mechanical testing can be seen in table 2. The yield ($Rp_{0.2}$) and tensile ($R_m$) strength of the materials are on a similar level but there are differences in elongation (A), area contraction (Z) and impact toughness (CVN). The A625:2 material has a bit high strength which most likely can be attributed to a higher C-content vs the A625:1 material as well as a higher Ti-content versus the nitrogen atomized materials. It is clear that elongation, area reduction and impact toughness are higher for the materials with higher oxygen content. This is contradictory to many other studies on Oxygen content in HIPed materials [1-6]. However comparing the results to what was found in the microstructure analysis it is clear that the materials with lower impact toughness has a very pronounced PPB structure decorated with oxides. It can be concluded that the oxide network on the PPBs have a very negative effect on the ductility and toughness of the material. This is confirmed when studying the fracture surfaces from impact toughness testing, figure 3a and 3b.

The fracture surface in the A625:1 material is characterized by high degree of PPB guidance of fracture, especially on larger powder particles but also on smaller. Some examples of this are pointed out with arrows in figure 3a. The uncovered powder particle PPB surfaces are fully covered with small dimples in which small alumina particles are found, often including elements of Ti. Both of which are strong oxide formers. In some such fracture surfaces, also larger particles enriched in Mo and Cr were observed however the alumina particles on the PPB completely dominate the fracture initiation and propagation in these materials.

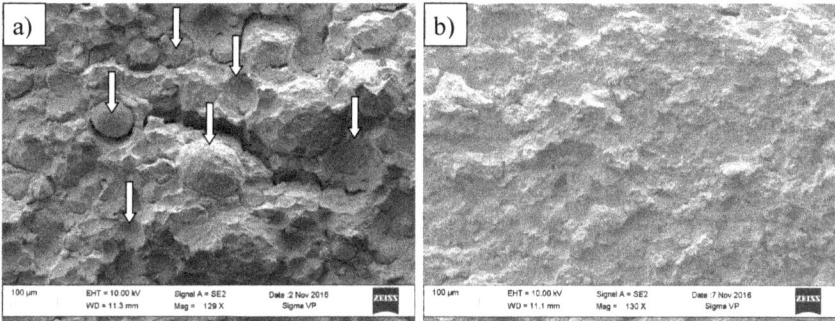

*Figure 3. Fracture surface of A625:1 (a) and A625:2 (b).*
*lectron Microscopy and Energy Dispersive Spectroscopy. portant plant for SPS to grow business in Ni-base products.*

The fracture in the higher O-content A625:2 is primarily transgranular at low magnification. Higher magnification reveals that the fracture is a combination of intergranular PPB guided fracture and transgranular fracture. Small dimples with alumina particles inside cover the uncovered PPB surfaces. NbC precipitates are also observed however at a much lower amount than the alumina (with Ti) oxides.

In the N625:1 material the fracture is almost purely PPB-guided fracture where the prior powder particles are uncovered by the impact test display dense occurrence of alumina particles, see figure 4a. The fracture in N625:2 is a combination of transgranular, guided by Nb, Cr, Mo rich nitrides and intergranular fractures guided by both nitrides and PPB-alumina particles, see figure 4b. The latter type of fracture appearance appears to be fairly uncommon but when it was clearly observed it appeared primarily on larger powder particles rather than on smaller powder particles.

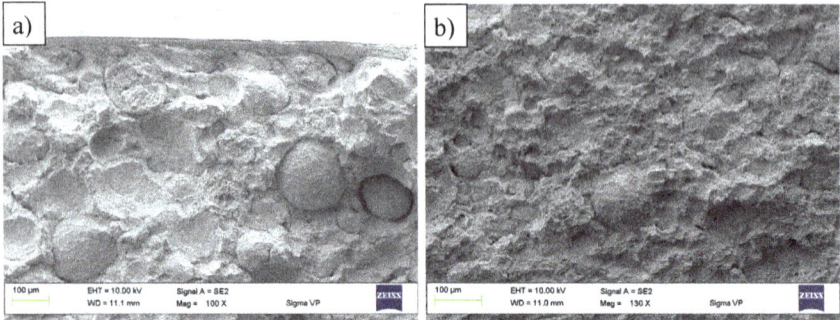

*Figure 4. Fracture surface of N625:1 (a) and N625:2 (b).*

**Oxygen distribution.** The results in the microstructural analysis combined with the results from mechanical testing clearly show that oxygen content alone cannot be used to verify that a material will perform as desired. The distribution of the oxygen in the powder as well as the surface after manufacturing and consequently the consolidated material is of importance. Figure 5 show the oxygen content in the materials for different powder size fractions. As expected the finer powder contains more oxygen, this due to a higher surface area. The black lines represent oxygen content in the melt prior to atomization i.e. bulk oxygen content which should also serve as a good indication of the bulk oxygen content in the powder. Making the assumption that the bulk oxygen content is the same for all size fractions, the oxygen uptake for each size fraction can be estimated.

The particle size distribution in the nitrogen atomized is almost identical. As can be seen in figure 5 there is a large difference in oxygen uptake in the two powders for all the size fractions. The low bulk oxygen in combination with high total oxygen content of the N625:1 material shows that most of the oxygen in that material is surface oxygen. This is also confirmed by the fracture surface analysis of the impact toughness specimens where it could be seen that the fracture was almost purely PPB guided fracture in the N625:1 material. The alumina particles found in the PPB guided fracture surfaces as little to no bond to the matrix around them and cracks can easily propagate along this path, effectively lowering the impact toughness.

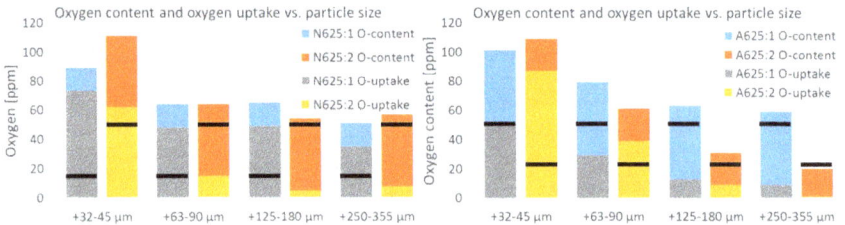

*Figure 5. Oxygen content for different particle size fractions for N-atomized material (left) and Ar-atomized material (right)*

Compared to the Nitrogen atomized material that had similar total oxygen content in all size fractions only the two finer fractions are on similar level while the coarser has a clear difference for the Argon atomized materials. The coarser powder in the A625:1 material has a significantly higher oxygen content compared to the A625:2 material although most of it is bulk oxygen. There is also a significant difference in oxygen uptake on the coarsest size fraction where the material with the lowest toughness also has the highest uptake, as with the nitrogen atomized powder, but not nearly as high. Also here the fracture in the impact toughness specimens is to the large extent PPB guided, primarily on the coarser particles but also on the finer.

## Conclusions

- Higher oxygen content material can have higher impact toughness
- Higher surface oxygen content on powder particles has much more adverse effect on impact toughness than bulk oxygen
- Fracture in material with a pronounced PPB network occurs along the PPBs resulting in lower impact toughness
- Fracture in material with lower surface and higher bulk oxygen content occurs trans granular resulting in higher impact toughness
- Results indicate that oxidation of coarse powder particles has a larger effect than oxidation of finer particles

## References

[1] A. Angré, A. Strondl, The effect of drastically lowered oxygen levels on impact strength for HIP'ed 316L-material, Preceedings HIP-14, 2014. p. 205-211.

[2] C. Cédat: Development of HIP technology on Stainless Steels, Paris, April 18th, 2013

[3] A. Lind, J. Sundström, A. Peacock, The effect of reduced oxygen content powder on the impact toughness of 316L powder joined to 316 steel by low temperature HIP, Fusion Engineering and Design, Vol 75-79, Nov 12-16 2000, Kyoto Japan.

[4] J. Sundström, Influence of oxygen content on impact strength of hot isostatically pressed PM superduplex stainless steel and 316L stainless steel, Swedish Institute for Metals Research, Report IM-2005-509 (2005), ISSN 1403-848X.

[5] J. Sundström, S. Wikman, S. Caddéo, Further assessment of the low oxygen pretreatment method for improved mechanical properties of hot isostatically pressed PM-alloys, Swedish Institute for Metals Research, Report IM-2005-554 (2005), ISSN 1403-848X.

[6] J. Sundström, The effect of oxygen on impact energy for HIPed PM-steels. Part 1, Swedish Institute for Metals Research, Report IM-2003-811 (2003)

[7] T. Berglund, L. Larsson, M Östlund, Impact toughness of HIPed vs. conventional 316L, ASMP PVP 2015, 45534

[8] T. Berglund, M. Östlund, Impact toughness for PM HIP 316L at cryogenic temperatures, ASMP PVP 2016, 64008. https://doi.org/10.1115/PVP2016-64002

Hot Isostatic Pressing – HIP'17                                    Materials Research Forum LLC
Materials Research Proceedings **10** (2019) 142-148          doi: http://dx.doi.org/10.21741/9781644900031-20

# Wear of PM HIP Metal Matrix Composites – Influence of Carbide Type

Tomas Berglund [1,a*], Josefine Hall [2,b]

[1] Sandvik Powder Solutions, PO Box 54, 79521 Surahammar, Sweden

[2] Dalarna University, 791 88 Falun, Sweden

[a]tomas.berglund@sandvik.com, [b]joh@du.se

**Keywords:** HIP, Wear, Tribology, Metal Matrix Composites, MMC, Wear Resistant

**Abstract.** The type of hard phase in combination with matrix material has a great influence on the wear properties of PM HIP Metal Matrix Composites. The hardness and toughness of the hard phase as well as its reaction with the matrix in combination with wear mechanism can cause significant differences in performance of the material. Three materials with the same matrix alloy but different carbide types have been studied with regards to tribological behavior in low stress and high stress abrasion as well as scratch testing against a quartz stylus.

In low stress abrasion testing the materials has only very small differences in the performance between the materials. The materials containing crushed or spherical fused tungsten carbide had a higher initial wear rate compared to the material with macrocrystalline carbide. This can be explained by the higher degree of carbide dissolution in these materials. In the later stages of wear the three materials have similar performance. In the scratch testing a clear difference can be observed between the materials. The material containing the fused tungsten carbide exhibits a higher degree of carbide damage at the exit side of the wear scar sliding over the carbide. This can be attributed to the much higher degree of carbide dissolution in the fused carbide compared to the MC carbide. The results from tribology testing are discussed and compared to wear mechanisms observed in parts that have been in service in a slurry pump and a crusher.

## Introduction

Wear-resistant materials are important in several industrial processes including manufacturing, energy production, construction and mining tools or in nuclear, aerospace and gas turbine industry [1-6]. Wear resistant metal matrix composites are composed of hard particles like carbides, borides and nitrides embedded in a ductile metal matrix of for example Co, Ni and/or Fe [2-5]. These two components serve different purposes. The hard particles need to impede abrasive wear by grooving or indenting hard particles while the metallic matrix should provide toughness. The properties of the composite material depend on both the chemistry of each component as well as the size and amount of the hard particles.

Wear resistant composites with tungsten carbide hard phase is commonly used as an overlay material in e.g. the mining industry. The overlay process used to apply these materials on to a part e.g. Plasma Transfer Arc Welding have limitations with regards to e.g. chemistry, hard phase size, overlay thickness etc. The use of Hot Isostatic Pressing provides an attractive alternative to these processes as it offers more freedom in the composition of the material and the wear resistant material can be applied to surfaces at an unlimited thickness.

The type and shape of tungsten carbide will affect the properties of the material. The fused tungsten carbides consisting of an approximate 80/20% mix of the $W_2C$ and WC phase respectively has a higher hardness than Macro Crystalline WC. However, the benefit of the MC carbide is that it resists dissolution during consolidation much better than the fused carbide, in which the $W_2C$ phase is more prone to dissolution [7]. The dissolution can affect the wear resistance of the material [8, 9]. Furthermore. the different tungsten carbide types have different mechanical properties that also affect the tribological behavior of the materials.

The effect of carbide type and morphology on the wear resistance of HIPed metal matrix composites were investigated using high and low stress abrasion testing as well as scratch testing. The resulting wear scars were investigated using SEM/EDS. The wear mechanisms found are discussed with respect to the microstructure and material properties.

## Experimental

**Materials.** The materials were produced at Sandvik Powder Solutions in Sweden using Hot Isostatic Pressing (HIP). The matrix and tungsten carbide powders were weighed and mixed to achieve a 50 vol.% carbide content in the material. The powders were mixed in V-blender for 3 hours. A NiCrSiBC alloy was used as matrix for all materials. The mixed powder were filled in to a capsule (180x70x50 mm) which was then evacuated and HIP:ed at 100 MPa and 1150°C with 2 hour dwell time. Three materials; with crushed fused tungsten carbide (80/20 mix of $W_2C$ and WC) (F50), spherical fused tungsten carbide (FSP50) and MC WC (MC) (F50MC) respectively. The carbide particle size used was 100-300μm. The same for all the materials although the particle size distribution vary slightly. From the HIPed specimens the samples were cut using Electric Discharge Machining.

Samples used in scratch testing were polished to 1 μm using diamond emulsion in the final step and were cleaned in acetone, water and ethanol before scratch testing. For abrasion testing the specimens were ground to remove white layer from EDM and to produce a controlled surface before testing. Resulting surface finish was approximately Ra 0,6 μm.

**Abrasion testing.** Low stress abrasion testing was performed using a modified version of procedure ASTM G65 A. When testing highly abrasion resistant materials it has been found that normally used one or two passes (see table 1 for test parameters per pass) is not enough to characterize the wear resistance of the material. This because after two passes the material removed in testing is only matrix material between the hard phase particles. Instead the testing has been made in 6 passes with weight loss measurement after each pass. The measured average weight loss (0,001 g precision) for three tests was then used to calculate volume loss.

High stress abrasion was carried out with the testing parameters seen in table 1. Average weight (0,01g precision) loss of three tests was used (0,01g precision). The basics of the test is described in standard ASTM B611 where it is also mentioned that it is the preferred test method for testing of cemented carbides.

**Hardness testing.** Hardness testing was performed using a CSM Micro Combi Tester for micro hardness measurement. Load was 200 g for carbide

*Table 1. Test parameters for low and high stress abrasion testing.*

|  | Low stress | High stress |
|---|---|---|
| Base procedure | ASTM G65 A mod. | ASTM B611 |
| Abrasive media | Silica sand #50-70 | 30# Al2O3 |
| HV abrasive | 1100 HV | 2200 HV |
| Slurry | N/A | 80% abrasive 20% H20 |
| Wheel material | Rubber | AISI 1020 Steel |
| Wheel RPM | 212 | 100±5 rpm |
| Sand flow rate | 375 g/min |  |
| Force | 130N | 200 N |
| No revolutions | 6000 | 1000 |
| Sliding distance | 4309 meters | 518-530 meter |

hardness and 25 g for matrix hardness. For macro hardness, HRC testing was performed in a

Hot Isostatic Pressing – HIP'17                                      Materials Research Forum LLC
Materials Research Proceedings **10** (2019) 142-148      doi: http://dx.doi.org/10.21741/9781644900031-20

Leco RT-240 hardness tester. The average was calculated from five tests. In hardness testing of carbides and matrix, indentations that caused significant cracking of the carbide or that contacted underling carbides (severe indent distortion) was excluded. Imprints on carbides were made on carbide particles that was at least ten times bigger than the imprint diagonal.

**Scratch testing.** A conventional scratch tester CSM Revetest® was used for scratching the sample surfaces. The stylus used was a pyramidal quartz ($SiO_2$) stylus. In the quartz scratch test the scratch length was 5 mm at the speed 10 mm/ min at the load 50 N.

## Results

**Microstructural characteristics.** The microstructure of the three materials is shown in Fig. 1. The F50MC material exhibits a very low amount carbide dissolution as well as a high degree of fracturing in these carbides. It is also evident that the angular eutectic carbide has a slightly higher degree of dissolution compared to the spherical. This is due to the slightly higher amount of the more dissolution prone $W_2C$ phase in the angular carbide as well as a higher surface area and coarser structure of the angular carbide. Left behind are needle shaped WC-phase extending out in to the matrix as well as smaller secondary precipitates in the matrix between the primary carbides. This dissolution increases the mean free path between the larger primary carbide particles that contributes mostly to the wear resistance of the material by resisting damage caused by abrasive particles. This can have a negative effect on the wear resistance of the material as the matrix between the primary carbides is more accessible to wear. The dissolution of the spherical carbide is much lower and a three-layer structure has formed in the carbide. A 2μm layer of what is believed to be MC carbide, a 10-15μm transition zone 2-phase WC/W2C structure but that has coarsened slightly and in the center the original structure of the spherical carbide.

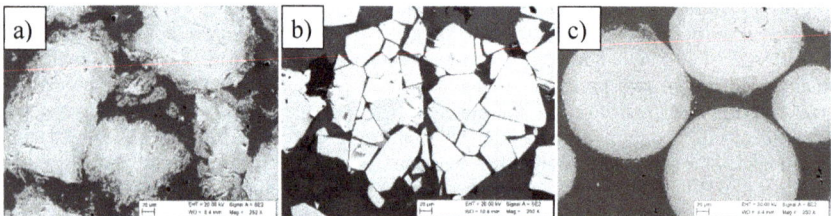

*Figure 1. Microstructure of F50 (a), MC50 (b) and FSP50 (c).*

**Mechanical characteristics.** The macro hardness of the materials as well as micro hardness of the matrix and carbides can be seen in figure 2. As can be seen the spherical eutectic carbide have the highest hardness followed by the fused crushed carbides and the macro crystalline carbide. A higher amount of carbide dissolution have increased the hardness of the matrix by solid solution strengthening by C and W but also precipitating small carbides in the matrix.

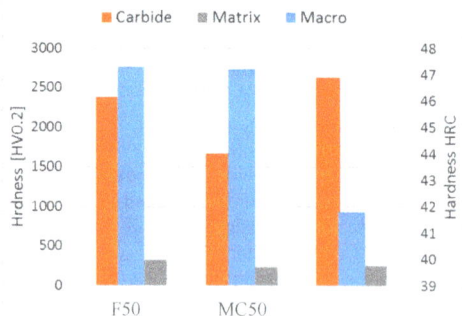

*Figure 2. Macro hardness of material and micro hardness of carbides and matrix.*

Hot Isostatic Pressing – HIP'17                                    Materials Research Forum LLC
Materials Research Proceedings **10** (2019) 142-148        doi: http://dx.doi.org/10.21741/9781644900031-20

**Tribological characteristics.** Figure 3 show the volume loss per pass and cumulative volume loss in low stress abrasion testing. It is evident that the F50MC material has a slightly lower initial wear compared to F50 and FSP50. This can be attributed to the lower degree of carbide dissolution in this material (see figure 1). Higher degree of carbide dissolution as in F50 increases the mean free path between the primary carbides causing abrading sand particles to more easily remove the softer matrix material along with the small carbides between the primary carbides. The higher degree of carbide fracturing during HIP in F50MC further lowers the mean

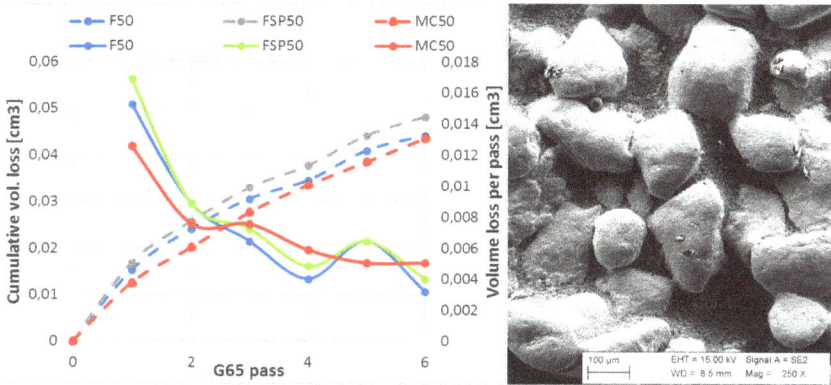

*Figure 3. Cumulative and volume loss per pass in cm³ in ASTM G65 testing and to the right the worn surface of F50 after 6ᵗʰ pass of testing.*

free path between the primary carbides, but also reduce their effective size. The FSP50 material with spherical particles performed worst, most likely a consequence of the spherical shape of the carbide particles combined with a size distribution towards the larger end increasing mean free path.

After the initial stage (pass 4) of wear, it's mostly matrix material that is being worn away, there is very little difference between the materials, and wear rate starts to stabilize. The hardness of the abrading particles is significantly lower than all the carbides. This in combination with the low contact pressure used results in that the carbides resists wear quite well. In fact, the material with the lowest hardness perform the best, although mainly due to better performance in the pass one and two. This indicates that during low stress abrasion the major controlling property is not hardness but rather mean free path of the carbides, the wear resistance of the matrix as well as the carbide-matrix bond strength. The controlling factor in the wear process is the removal of the matrix material between the carbides. When a sufficient amount of matrix is worn away, the carbide particles

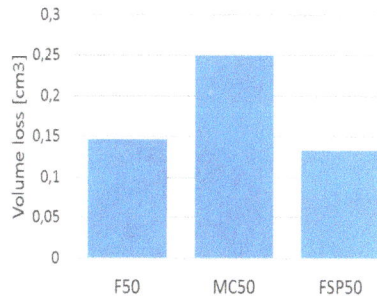

*Figure 4. Average volume loss in high stress abrasion testing after 1000 revolutions.*

lose their support from the matrix and can readily be removed as wear progresses. As can be seen in figure 3 the wear rate in the material does not start to stabilize until the 4th to 6th pass of testing. Even at this stage the total depth of the wear scar is still quite low relative to the carbide size and the majority of the wear has occurred in the matrix between the carbide particles, see figure 3.

Figure 4 shows the average vol. loss in high stress abrasion testing. The harder angular abrasive particles in combination with a higher load and the harder counter surface leads to much more severe wear conditions and significantly higher vol. loss. In this test the abrasive particles generally fracture during the test and other studies have shown that the fractured particles tend to produce even greater wear than the originally added particles [10].

A clear correlation can be found between carbide hardness and wear resistance of the materials. It is evident that the softer MC carbide in the F50 MC material results in a lower wear resistance in this test. It has been shown in other studies [11] that the abrasive wear rate drops drastically when a materials hardness goes above 80% of the abrading particles. For the low stress abrasion testing the hardness of the carbides for all the materials are more than twice that of the abrading particles. However, for the $Al_2O_3$ particles used in the high stress abrasion testing the hardness of macro crystalline carbide used in F50 is only about 75% of the abrading particles. For F50 and FSP50 the corresponding values are 110% and 120% respectively.

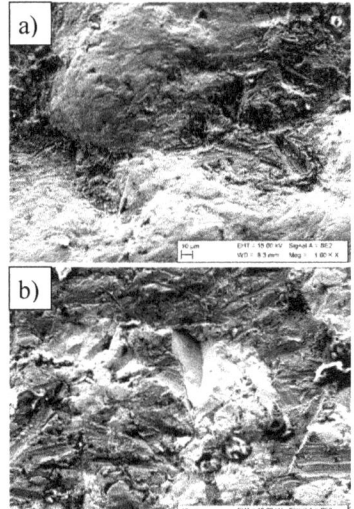

Figure 4. Surface of F50 (a) and MC50 (b) after high stress abrasion

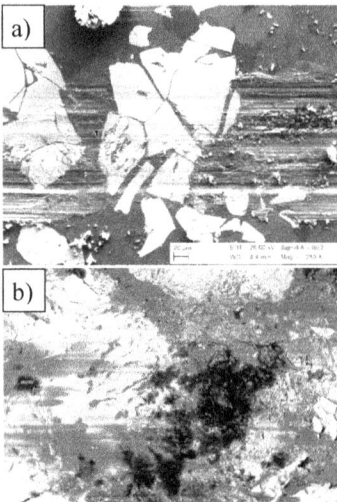

*Figure 5. Surface of MC50 (a) and F50 (b) after scratch testing against a*

Studying the worn surface on F50MC from high stress abrasion testing (figure 5) it becomes clearer why it performs slightly worse than the others. The carbides in the F50MC material exhibit more fracturing compared to the fused carbides, that show very little signs of fracturing. In F50 the wear is dominated by rounding of the carbide particles followed by a gradual material removal (figure 5). There is a marked difference in topography in the wear scar comparing the materials. In the material containing the harder fused carbides the carbides protrude from the surface, a clear indication that they quite effective resisting the wear from the $Al_2O_3$ particles. The topography in the wear scar for F50MC material is much smoother and contains more debris in the form of carbide fragments but also $Al_2O_3$ particles.

In scratch testing the MC carbide again displays a

Hot Isostatic Pressing – HIP'17                                    Materials Research Forum LLC
Materials Research Proceedings **10** (2019) 142-148          doi: http://dx.doi.org/10.21741/9781644900031-20

tendency to crack more compared to the fused carbides, see figure 6. Even though the MC carbide has a lower dissolution most of the positive effect gained from this is lost due to the susceptibility to cracking as well as a lower hardness.

The fused carbide exhibit significant damage on the edges where carbide dissolution has formed a structure that readily deforms and cracks i.e. it increases the mean free path between the more solid core of the primary carbides. Comparing this to the results in abrasion testing, it is apparent that this dissolution accelerates the wear in the matrix as well as the dissolution zone between matrix and primary carbide.

The worn surfaces from lab testing have been compared to worn surfaces of parts that have been in service under different operating condition. One where much of the wear is closer to low stress abrasion (pump part) and one that is subjected to high stress abrasion and impacts (crusher tooth). Both parts are using a wear material that is similar to F50 but have an optimized matrix composition. There are some clear similarities to the lab scale testing when studying the worn surfaces. For the crusher tooth the worn surface is quite flat with lots of fractured carbides (figure 7 a), quite like what was observed in the high stress abrasion testing. The depth in to the material to which the carbides are fractured is surprisingly low for it being an application with impacts wear also, only about 0,5 mm.

Studying the worn surface of the pump part instead (figure 7b) the similarity to the worn surface from the low stress abrasion test is significant. The carbides stand proud from the surface and the matrix between them is worn away. Unlike the surface from lab testing the surface contain some carbide damage like abrasion, cracking and micro chipping, although to a quite low extent.

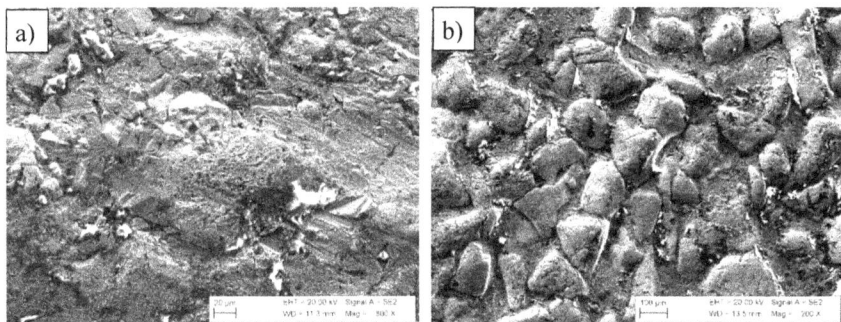

*Figure 6. Worn surface of crusher tooth (a) and pump side liner (b).*

**Conclusions**

- Angular fused tungsten carbide exhibit a higher degree of carbide dissolution compared to spherical fused carbide when HIPed at the same conditions
- Macrocrystalline carbide exhibits lowest degree of carbide dissolution compared to both of the fused carbides
- Lower carbide dissolution result in better wear resistance in the early stages of wear during low stress abrasion testing
- There is a good correlation between carbide hardness and wear resistance in high stress abrasion testing
- Higher degree of carbide dissolution increases the hardness of the matrix

- The wear mechanism in the pump part is similar to low stress abrasion
- The wear mechanism in the crusher tooth is similar to high stress abrasion

**References**

[1] Y.Pan, D. Y. Li and H. Zhang, Wear 271 (2011) 1916-19-21.

[2] H. Berns, Wear 254 (2003) 47-54.

[3] S. Bodhak, B. Basu, T. Venkateswaran, W. Jo, K.-H. Jung and D.-Y. Kim, J. Am. Ceram. Soc. 89[5] (2006) 1639-1651.

[4] R. M. Genga, G. Akdogan, J.E. Westraadt and L.A. Cornish, Int. J. Refract. Met. Hard Mater. 49 (2015) 240-248.

[5] R. M. Genga, G. Akdogan, C. Polese, J.C. Garrett and L.A. Cornish, Int. J. Refract. Met. Hard Mater. ARTICLE IN PRESS (2014).

[6] S. Olovsjö, R. Johanson, F. Falsafi, U. Bexell and M. Olsson, Wear 302 (2013) 1546-1554.

[7] Seger. R, Effect of tungsten carbides properties of overlay welded WC/NiSiB composite coatings, Thesis work, 2013, Höganäs

[8] T. Liyanagea, G. Fisher, A.P. Gerlicha, Wear 274-275 (2012) 345-354

[9] M. Jones, U. Waag, Wear 271 (2011) 1314-1324

[10] S. Wirojanupatump, P.H. Shipway, Wear 239 (2000) 90-101

[11] Jacobson, Staffan & Hogmark, Sture (1996). Tribologi: friktion, smorjning och nötning. 1. utg. Stockholm: Liber utbildning

Materials Research Proceedings 10 (2019) 149-156

doi: http://dx.doi.org/10.21741/9781644900031-21

# High Pressure Heat Treatment - Phase Transformation under Isostatic Pressure in HIP

Magnus Ahlfors[1,a] *, Alexander Angré[2,b], Dimitris Chasoglou[3,c], Linn Larsson[4,d]

[1]Quintus Technologies LLC, Lewis Center, OH, US

[2]Carpenter Powder Products AB, Torshälla, Sweden

[3]Höganäs AB, Höganäs, Sweden

[4]AB Sandvik Materials Technology, Sandviken, Sweden

[a]magnus.ahlfors@quintusteam.com, [b]aangre@cartech.com, [c]dimitris.chasoglou@hoganas.com, [d]linn.larsson@sandvik.com

Keywords: HIP, High Pressure Heat Treatment, URQ

**Abstract.** Modern HIP furnaces equipped with forced convection cooling enable very fast cooling rates under isostatic pressure. This does not only give shorter HIP cycles and increased productivity but also allows complete heat treatment cycles to be performed in the HIP unit. It has been shown in previous studies that extreme pressures of several thousand bar can push phase transformation towards longer times for the Fe-C system. The new URQ HIP cooling systems give the opportunity to investigate the impact of pressures up to 2000 bar on phase transformation time dependency. A 4340 steel was used in this study and a comparison of austenite phase transformation time at 100 bar and 1700 bar was performed. The study was performed by isothermal heat treatment of specimens for a specific time followed by quenching. To evaluate the influence of pressure on hardenability, the phase fractions were evaluated using grid method on SEM images. The study found significant influence of HIP pressure on the phase transformation kinetics of the material studied.

## Introduction

Hot Isostatic Pressing (HIP) is a process mainly used to consolidate powder into solid high-quality parts or to eliminate internal defects in parts produced by casting, additive manufacturing or MIM by applying a high isostatic gas pressure and a high temperature. Traditionally the cooling in the HIP system is relatively slow and could take up to 24 hours. In the mid 1980's the URC HIP furnaces was introduced with a forced convection cooling technology that significantly decreased the cooling time in the HIP system and thereby reduce the total HIP cycle time by up to 50% [1]. In 2010 the URQ HIP furnaces were introduced with achievable cooling rates up to 3000 K/min. The URQ HIP quenching furnaces gives the possibility to perform traditional heat treatments, e.g. martensitic hardening, in the HIP furnace during the HIP cycle.

The forced convection cooling technology (URC, URQ) is based on a wire wound pressure vessel design where a thin cylinder is water cooled from the outside and a wire wound package outside the cooling channels. To protect the pressure vessel from heat during a HIP cycle, an insulated furnace within the pressure vessel is used to achieve high temperature for the load in the hot zone but a cool environment closest to the pressure vessel walls. During the forced convection cooling the hot gas inside the furnace is moved to the outside of the furnace at the same time as the colder gas outside the hot zone is pushed into the furnace chamber. This mixing of gas will lead to a cooling effect and at the same time the hot gas outside the furnace is cooled down by the water-cooled pressure vessel walls like a heat exchanger which adds to the cooling

Materials Research Forum LLC

doi: http://dx.doi.org/10.21741/9781644900031-21

effect. In the case of a URQ furnace a heat exchanger is also placed inside the pressure vessel, outside the furnace, to increase the cooling rate even more. In Figure 1, a schematic image of a URQ furnace during cooling is presented.

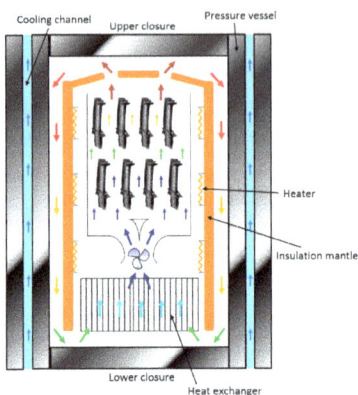

*Figure 1. Schematic image of the cooling in a URQ HIP furnace*

When looking at performing quenching of a material inside the HIP during a HIP cycle the question came up if the high isostatic pressure on the material during the quenching would make any difference from the conventional quenching methods at atmospheric pressure.

A few relatively old geological studies have shown that high pressure during cooling pushes the phase transformation from austenite to pearlite and bainite towards longer times. Most studies have been evaluating very high pressures; 20-42 kbar. Kuteliya et al. studied the effect of 20 kbar hydrostatic pressure on austenite transformation and saw that high pressure slows down the transformation of austenite. [7] In a study made by Radcliffe et al. it was shown that both the initiation and the rate of austenite transformation in iron-carbon alloys are retarded at 42 kbar, relative to the reactions at 1 atm. [8] Austenite – pearlite transformation rates at 34 kbar pressure were studied by Hilliard et al. and it was found that the effect of the pressure was a decrease in transformation rate. [9]

A.Weddeling has used the rapid quenching HIP technology to study the influence of high pressure on microstructure. Specimens of low alloyed steel were quenched under pressure in HIP and compared to samples quenched with the same cooling rate in a dilatometer at atmospheric pressure. A certain amount of bainite was formed during quenching but the specimen quenched in the HIP featured less bainite that also had finer structure compared to the dilatometer specimens. These results suggest that the bainite formed under pressure is not only formed later in time but also at lower temperatures at which nucleation is supported and diffusion is retarded compared to bainite formation at higher temperatures. [6]

The object of this study was to investigate if a typical HIP pressure of 1700 bar is enough to shift the austenite to pearlite phase transformation towards longer times and if so of what order of magnitude. Shifting the phase transformation towards longer times would imply an increased hardenability which could be very beneficial in industry.

For this study, two different steels were studied. For each material a comparison of austenite to pearlite phase transformation time at 100 bar and 1700 bar was performed. The study was performed by isothermal heat treatment of specimens for a specific time followed by quenching. To evaluate the influence of pressure on the hardenability, the phase fractions were evaluated by grid method in SEM together with hardness measurements. See Figure 2 for a schematic presentation of the thermal profiles of the different HIP cycles.

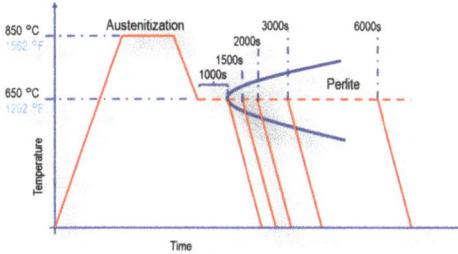

*Figure 2. Schematic presentation of experimental HIP cycles*

## Experimental
*Material*
*4340*
Steel 4340, as well denominated 34CrNiMo6 (EN name) and SS 2541, is a widely used steel for quenching and tempering. The chemical composition of the 4340 material used in the study is displayed in Table 1.

*Table 1. Chemical composition of the 4340 material according to specification.*

| Elements (%) | C | Mn | Si | Cr | Ni | Mo | Cu | Fe |
|---|---|---|---|---|---|---|---|---|
| 4340 | 0.37 | 0.74 | 0.26 | 1.45 | 1.50 | 0.19 | 0.16 | bal. |

*TTT diagrams*
The 4340 material was chosen among several similar steels consulting their Time Temperature Transformation (TTT) diagrams. The factors in favour for 4340 were that it was a fairly common material and that the time to pearlite start was sufficient for the experiments. TTT diagrams for 4340 were found in literature and educational material and additionally JMatPro was used to calculate a diagram. As can be seen in Figure 3 and 4, the TTT diagrams vary quite a bit among themselves and this had to be taken into account deciding isothermal hold time intervals.

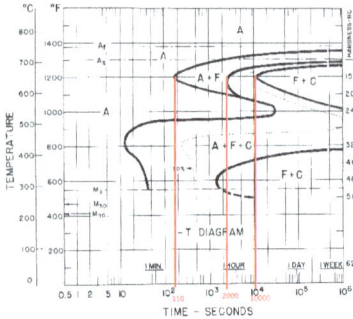

*Figure 3. TTT diagram for 4340 [10].*
*1% ferrite at 150 s, 1% pearlite at*
*2000 s and 99% pearlite at 10000 s at*
*650 °C.*

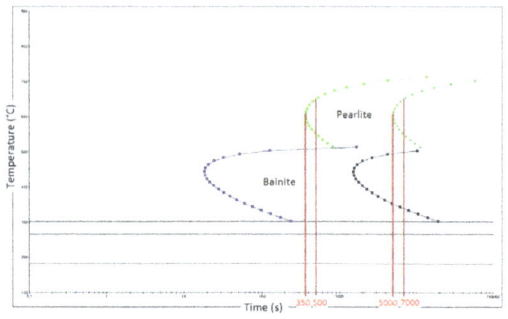

*Figure 4. TTT diagram for 4340 with grain size 15 μm,*
*calculated using JMatPro. 0.1% pearlite at 500 s and*
*99.9% pearlite at 7000 s at 650 °C.*

## Trials

Suitable isothermal temperature was selected consulting available TTT diagrams for the material. Subsequently, suitable hold time for the material was also selected based on the TTT diagrams. The HIP cycle is designed so that the material initially is subjected to austenitization temperature, 850 °C, for 15 minutes followed by fast cooling down to selected isothermal temperature, 650 °C, where the material is kept for the chosen hold time followed by rapid quenching to room temperature. The average time to quench the material from 850 to 650 °C in the HIP was 35 seconds measured with the TC in the sample. All HIP cycles were performed at low and high pressure separately. The main challenge for running the HIP cycles was to achieve the same thermal profile for the materials for the two different pressures.

All HIP trials were performed with solid cylindrical specimens with size 25x25 mm. Thermocouples were placed in the gas of the HIP furnace hot zone and at least one thermocouple in the center of the specimen for all HIP trials, see Figure 6. The samples were prepared with holes, drilled halfway through the height of the sample in which the thermocouples were placed in order to measure the temperature in the center of the sample. All HIP cycles were performed in the QIH9 URQ HIP at Quintus Technologies AB, Västerås, Sweden, permitting cooling rates of up to 3000 K/min in the gas, i.e. about 45 K/s.

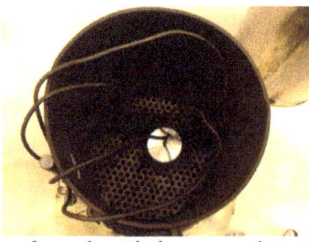

*Figure 6. Set up of sample with thermocouple in the HIP furnace.*

Hot Isostatic Pressing – HIP'17                                   Materials Research Forum LLC
Materials Research Proceedings **10** (2019) 149-156        doi: http://dx.doi.org/10.21741/9781644900031-21

Table 3 shows the test matrix of all the trials performed within the project. Typical HIP log curves are presented in Figures 7 and 8, displaying the HIP log curves for 4340 with 3000 s hold time at low and high pressure respectively.

The maximum pressure possible to run the rapid quenching feature in the used HIP system is 1700 bar so that pressure was used for the high-pressure cycles of this study. The low-pressure cycles were decided to be run in the HIP as well, in order to make the high and low pressure thermal profiles as similar as possible. To be able to control the HIP temperature and rapidly quench the material 100 bar had to be used as minimum pressure.

*Table 3. Matrix of the trials performed within the project.*

| Sample | Austenitization temperature [°C] | Austenitization time [min] | Isotherm [°C] | Pressure [bar] | Hold time [s] |
|---|---|---|---|---|---|
| 4340 1000s | 850 | 15 | 650 | 100 | 1000 |
| 4340 1000s P | 850 | 15 | 650 | 1600 | 1000 |
| 4340 1500s | 850 | 15 | 650 | 100 | 1500 |
| 4340 1500s P | 850 | 15 | 650 | 1700 | 1500 |
| 4340 2000s | 850 | 15 | 650 | 100 | 2000 |
| 4340 2000s P | 850 | 15 | 650 | 1700 | 2000 |
| 4340 3000s | 850 | 15 | 650 | 100 | 3000 |
| 4340 3000s P | 850 | 15 | 650 | 1700 | 3000 |
| 4340 6000s | 850 | 15 | 650 | 100 | 6000 |
| 4340 6000s P | 850 | 15 | 650 | 1800 | 6000 |

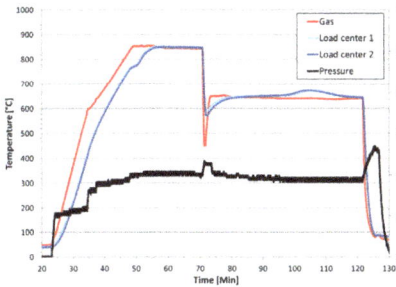

*Figure 7. HIP log curve for 4340, 3000 s, 100 bar.*

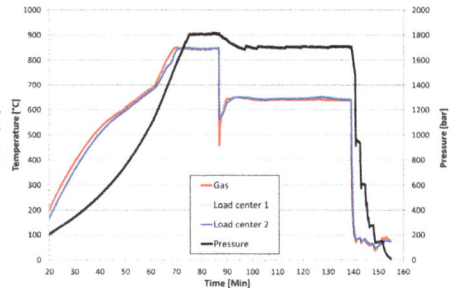

*Figure 8. HIP log curve for 4340, 3000 s, 1700 bar.*

*Phase fraction evaluation*

The phase fraction evaluation was performed at Swerea KIMAB AB, Kista, Sweden. Prior to evaluation, each specimen was wet ground, polished with diamond paste to 0.25 μm and subsequently electrolytically polished (20V, 20s). The phase fraction evaluation was performed using the grid method on SEM images. The FEG-SEM equipment used in this work was a LEO

1530 with Gemini column, upgraded to a Zeiss Supra 55 (equivalent). The SEM settings used was 5 kV with aperture 30 μm and the InLens detector. The micro Vickers hardness testing was performed in a Qness Q10 A+ hardness tester with a load of 1 kg.

### Results

Figure 9 and 10 display the results for 4340 in form of comparisons of phase amount transformed to pearlite and hardness, at low and high pressure, respectively. The graphs displaying the equivalent results for Astaloy Mo are shown in Figure 11 and 12.

To illustrate the pearlitic/martensitic microstructures of the samples, SEM images for 4340 with 2000 s hold time at low and high temperature are displayed in Figure 13 and 14 respectively.

For all the HIP log curves with longer isothermal times a "bump" in temperature can be seen during the 650 °C isotherm for the temperature measured inside the specimen although the temperature in the gas stays constant. This is most likely representing the exothermal austenite to pearlite transformation. See Figure 6 for example.

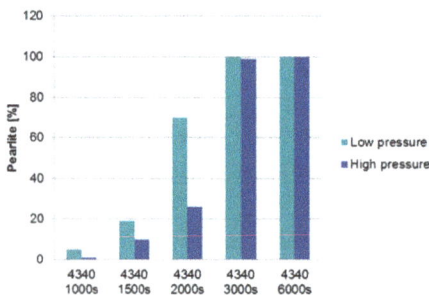

*Figure 9. Comparison of phase amount transformed to pearlite, at low/high pressure, for each of the 4340 samples.*

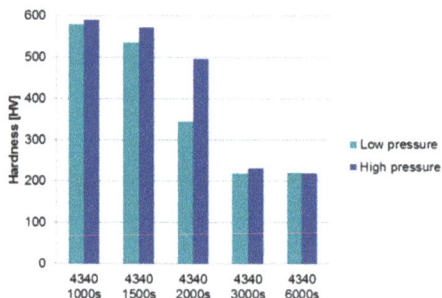

*Figure 10. Comparison of hardness, at low/high pressure, for each of the 4340 samples.*

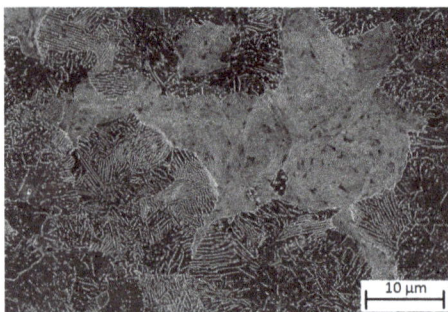

*Figure 13. SEM images of 4340, 2000 s, 100 bar. SEM magnification setting x5000.*

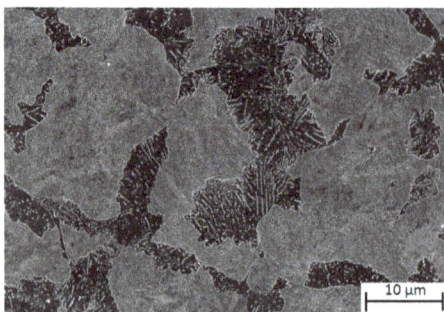

*Figure 14. SEM images of 4340, 2000 s, 1700 bar. SEM magnification setting x5000*

**Discussion**
The phase fraction evaluations unanimously show that HIP pressures influence the phase transformation from austenite to pearlite by pushing it towards longer times. For 4340 the 1500 s and 2000 s samples give the clearest results where the 2000 s sample subjected to low pressure contains almost three times as much pearlite as the sample subjected to high pressure. The 1000 s and 3000 s samples indicate the same influence but to a less numerical extent due to low and high transformation degree respectively. The transformation degree is even greater for the 6000 s sample, giving 100% transformation at both pressure levels. Thus no information regarding influence of pressure can be drawn from the 6000 s sample. The hardness testing supports this conclusion for all samples. The hardness values are higher for the high pressure samples due to less pearlite phase transformation, i.e. higher amounts of martensite. For Astaloy Mo the phase transformation evaluation and the hardness measurements for both HIP cycles, 500 s and 1000 s, show that HIP pressure increases hardenability.

The 4340 HIP cycle logs display an increase in the load center temperature during the isotherm at 650 °C, although the gas temperature stays constant, most likely representing the exothermic austenite to pearlite phase transformation. As can be seen in the HIP cycle logs, the bump appears both larger and earlier in time at low pressure than it appears at high pressure. Not only does this phenomenon strengthen the thesis that HIP pressure pushes the austenite – pearlite phase transformation towards longer times but it also suggests that the pressure suppresses the rate of the phase transformation.

It is interesting to note that the results from the HIP trials are quite far from the TTT-diagrams found in literature and calculated using JMatPro, particularly for 4340. The results suggest that the 4340 material is fully pearlitic after 3000 seconds at 650 °C at 100 bar. This is quite far from the 5000 to 10000 seconds proposed by the TTT diagrams at atmospheric pressure. According to the thesis and the obtained results, the 100 bar pressure would, if anything, push the pearlite transformation towards longer times, not shorter. If the isothermal temperature of the HIP trials differed from the actual transformation peak temperature, this would also have resulted in longer time periods for pearlite phase transformation, and not shorter. What stands out for Astaloy Mo is that the material is fully martensitic after 500 s already. Possibly, an explanation to this is that the temperature, during the 500 s HIP trial, went quite low during the drop from 850 °C and then never reached 650 °C for the isotherm before final quenching. However, the TTT diagrams were used as base for selecting the experimental parameters. To determine exact time and temperature limits for phase transformation was not the scope of this project. For that, dilatometer trials with smaller specimens are more suitable.

These studies show that an increased hardenability for two steels is achieved when performing the quenching under pressure in an URQ HIP for examples. This is an interesting fact for industry since increasing a material's hardenability often is a very positive thing. For example, less alloying elements can be used to make a leaner and cheaper material. Thicker sections for a specific material that weren't possible to through harden before can now be just that. For distortion and crack sensitive components the quenching rate can be decreased to decrease the risk of cracking and distortion but still achieve the same quench effect on the microstructure.

**Conclusions**
All phase fraction evaluation results support the theory that HIP pressure pushes austenite – pearlite transformation towards longer times.

All hardness measurement results support the theory that HIP pressure pushes austenite – pearlite transformation towards longer times.

The bump in the temperature HIP log during the isotherm, representing the austenite – pearlite transformation, indicates that HIP pressure pushes austenite – pearlite transformation towards longer times and also that the HIP pressure suppresses the intensity of the austenite – pearlite phase transformation.

**Acknowledgments**

This work has been financed by the member companies of the Powder Materials Member Research Consortia at Swerea KIMAB and RISE. The members of the research committee of the project:

**References**

[1] A. Åkerberg, The Difference Between URQ and U2RC, Proceedings HIP'14 11th International Conference on Hot Isostatic Pressing, Stockholm, Sweden, June 2014.

[2]. A. Åkerberg, P. Östlund, Numerical Calculations of New Innovative Heat Treatment, Proceedings from HIP'14 the 11th International Conference on Hot Isostatic Pressing, Stockholm, Sweden, June 2014.

[3]. A. Åkerberg, Temperature Accuracy in HIP Furnaces, Proceedings from HIP'14 the 11th International Conference on Hot Isostatic Pressing, Stockholm, Sweden, June 2014.

[4]. A. Ahlfors, The Possibilities and Advantages with Heat Treatments in HIP, Proceedings from HIP'14 the 11th International Conference on Hot Isostatic Pressing, Stockholm, Sweden, June 2014.

[5]. R. Larker, P. Rubin, Uniform Rapid Quenching Enables Austempering Heat Treatment in HIP, Proceedings from the 6th International Quenching and Control of Distortion Conference, Chicago, USA, September 2012.

[6]. A. Weddeling, N. Wulbieter, W. Theisen, Densifying and Hardening of Martensitic Steel Powders in HIP Units Providing High Cooling Rate, Proceedings from EuroPM 2015, Reims, France, October 2015.

[7]. É.R. Kuteliya, L.S. Pankratova and É.I. Estrin, Isothermal Transformation of Austenite under High Pressure, Translated from Metollovedenie i Termicheskaya Obrabotka Metallov, No. 9, pp. 8-13, September, 1970.

[8] S.V. Radcliffe, M. Schatz and S.A. Kulin, The Effect of High Pressure on the Isothermal Transformation of Austenite in Iron-Carbon Alloys, Journal of the Iron and Steel Institute, pp143-153, February 1963.

[9] J.E. Hilliard and J.W. Cahn, The Effect of High Pressure on Transformation Rates, Progress in very high pressure research 1961, pp. 109-125, General Electric Research Laboratory, USA.

[10] American Society for Metals, Atlas of Isothermal and Cooling Transformation Diagrams, Metals Park Ohio, USA, 1977.

[11] http://personal.teknik.uu.se/hugon/Konstruktionsmaterial/Konstr_mat_HT2006/Laborationsi nstruktion_H%C3%A4rdning.pdf

Hot Isostatic Pressing – HIP'17            Materials Research Forum LLC
Materials Research Proceedings **10** (2019) 157-168     doi: http://dx.doi.org/10.21741/9781644900031-22

# Controlled Uniform Load Cooling in
# Production Scale HIP Equipment

Beat Hofer[1,a *], Michael Hamentgen[2,b], Maxime Pauwels[3,c],
Carlo Verbraeken[3,d], Pierre Colman[3,e]

[1]Steinmattstrasse 25, CH4552-Derendingen, Switzerland

[2]Werner-von-Siemens-Straße 23, D-66793 Saarwellingen, Germany

[3]Walgoedstraat 19, B-9140 Temse, Belgium

[a]beat.hofer@hoferwmb.ch, [b]michael.hamentgen@saar-pulvermetall.de,
[c]maxime.pauwels@epsi.be, [d]carlo.verbraeken@epsi.be, [e]pierre.colman@epsi.be

**Keywords:** HIP, SPM, EPSI, High Pressure Vessel, Frame, Heatshield, Radial Furnace, Mo-Heater, Thermocouple, Load, Temperature Accuracy, Cooling Parameter, Natural Cooling, Rapid Cooling, Jet Cooling, Controlled Cooling, Cooling Rate, Temperature Accuracies +/- 5°C

**Abstract.** A HIP system with a useful volume of Ø 800 x 2500 mm will be presented. With the help of jet cooling, a cooling rate of 40°K/min. is achieved in the range of 1220°C to 850°C. The uniformity of temperature distribution is between +/- 5°K during cooling. It is possible to drive different cooling rates. The function of the system is described.

## Introduction

Since April 2017, a second HIP unit with jet cooling, delivered and installed by Engineered Pressure Systems International NV (EPSI), is located at Saar-Pulvermetall GmbH in Saarwellingen, Germany.

This paper introduces the different cooling possibilities and demonstrates realistic cooling values in a large production size Hot Isostatic Press.

## Machine overview

The main components of the high pressure vessel (cylinder and closures) are forged from a vacuum degassed piece of steel. This material, a further development of AISI 4340 steel, corresponds to the requirements of the ASME-SA-723 specification.

The yoke frame structure takes up the axial forces on the closures that are generated by the pressure in the high pressure vessel. The yoke frame consists of 6 steel plates. Each plate is cut out of one rolled steel plate with no welded connections and is mounted to the framework.

The upper closure of the high pressure vessel is lifted by two hydraulically actuated cylinders. The high pressure vessel is locked and unlocked by a hydraulically actuated sliding plate and prevents the vessel from opening at a pressure smaller than 0.5 bar. The high pressure assembly is tested at 1.57 x the working pressure or 220 MPa. See figure 1 of the unit.

Materials Research Proceedings **10** (2019) 157-168       doi: http://dx.doi.org/10.21741/9781644900031-22

*Figure 1 – HIP140-880\*2500M*

The furnace module consists of 3 main components:
1. Thermal heatshield
2. Complete radial furnace with plug-in connections for thermocouples and power
3. Complete bottom furnace with baseplate and plug-in connections for thermocouples and power

The Mo-furnace has a useful diameter of 800 mm, a useful length of 2500 mm and a useful volume of 1256 dm$^3$. The maximum working pressure is 140 MPa. The maximum temperature is 1400°C.

The heating system is divided into 6 separate radial heating elements and 2 bottom heating elements in order to achieve a precise temperature distribution. Each heating zone has 2 regulating thermocouples. The first thermocouple serves as a control and monitoring input and the second one is used as a stand-in, in the event of any failure or interruption of the first one. In addition, it is possible to mount 10 loading thermocouples that are distributed along the entire furnace volume. When required, these loading thermocouples can be used as a control input.

**Requirements for temperature accuracy and cooling parameters on new HIP applications**
The ranges of application for hot isostatic presses are broadly diversified. They include the isostatic densification of castings, ceramics and hard metal parts. In the last few years, the elimination of pores in injection moulded parts and also increasingly the additive manufactured parts were added in this area.

Powder Metallurgy together with near-net-shape techniques enable the manufacturing of compact components by means of a capsule. This also includes the production of material combinations using different materials. All these materials have to comply with special requirements regarding tensile strength.

This also leads to different specifications with regard to HIP parameters. Various acceptance criteria were defined by the end-user in order to meet these demands as comprehensively as possible:

- Heating up to 1220 °C by 5 K/minute minimum
- Cooling down to 800 °C by 40 K/minute minimum
- Temperature accuracy during the entire heating, dwell and cooling period must be +/- 7°C at temperatures higher than >800°C and a pressure of >10 MPa.
- The criteria are specified on a loading weight of 2500 kg and a load carrier of 300 kg, both made of regular steel.

**Technology basis for reaching the desired parameters**
Heating is done through the radial and bottom heaters heating up the gas in the vessel. The efficiency of the heating depends on how well the heat shield keeps the heat in the hot zone.

*Natural cooling* takes place in its most basic form by switching off the heaters. As a result, a natural cooling curve can be expected in which the hot zone cools down by means of radiation losses through the heat shield to the cold vessel wall.

These losses are minimal (must also be the case, otherwise the heat shield is not well designed), as a result, this can take quite a long time.

The natural cooling can be accelerated by relying also on convection. The gas in the hot zone must go outside the heat shield to come into contact with the cooled vessel wall. This is called *rapid cooling.*

*Jet cooling* consists of using the compressor to pump cold gas into the hot zone and thus to achieve even better circulation and convective heat dissipation.

In addition to the above-mentioned cooling options, there is the possibility to glow with the heaters in order to neutralize temperature non-uniformities. We call this *controlled cooling*.

The cooling possibilities have considerably improved. They can now also be used for material treatment in large production HIPs and not only for shorter cycle times. This shows the machine at Saar Pulvermetall.

*Natural cooling*
In this variant, the heaters are switched off and the HIP cools slowly from bottom to top. Because the top and bottom are at different temperatures throughout the cooling cycle, it can be expected that there will be stresses in the material. This method is thus only used for applications where only the porosity of the material has to be remedied, but where residual stresses do not pose a significant problem. See the graph (figure 2) below for expected temperature distribution using this process.

Hot Isostatic Pressing – HIP'17                                    Materials Research Forum LLC
Materials Research Proceedings **10** (2019) 157-168          doi: http://dx.doi.org/10.21741/9781644900031-22

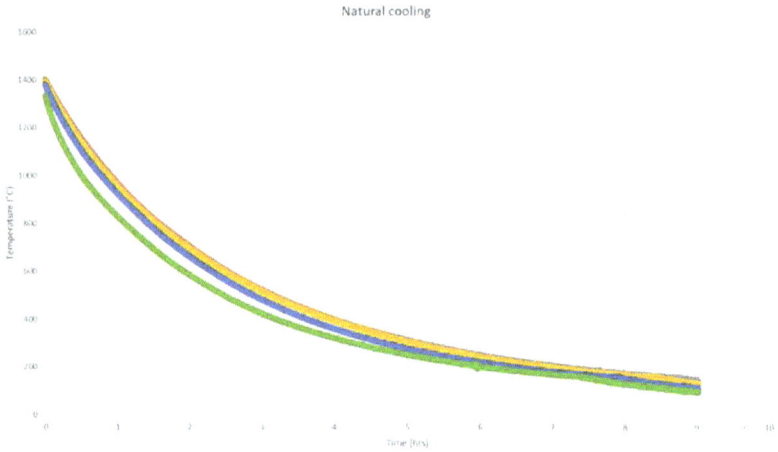

*Figure 2 – Natural cooling*

*Rapid cooling*
Rapid cooling is a first step to achieve faster cycle times. Rapid cooling means opening the lower valve to allow a circulation of gas in the hot zone outside the heat shield, against the vessel wall and back in the hot zone. See the figure 3 below for clarification.

Hot Isostatic Pressing – HIP'17                                    Materials Research Forum LLC
Materials Research Proceedings **10** (2019) 157-168          doi: http://dx.doi.org/10.21741/9781644900031-22

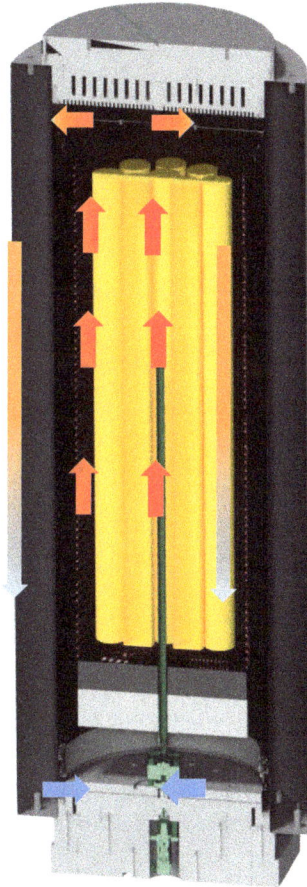

*Figure 3 – Rapid cooling schematic*

To get this speed high, it is important to obtain a good convective flow and also an efficient way to cool this flow.

It is important to mention that with this method the cooling speed is already considerably improved, but that also a non-uniform temperature distribution is the result. See curve in figure 4 below.

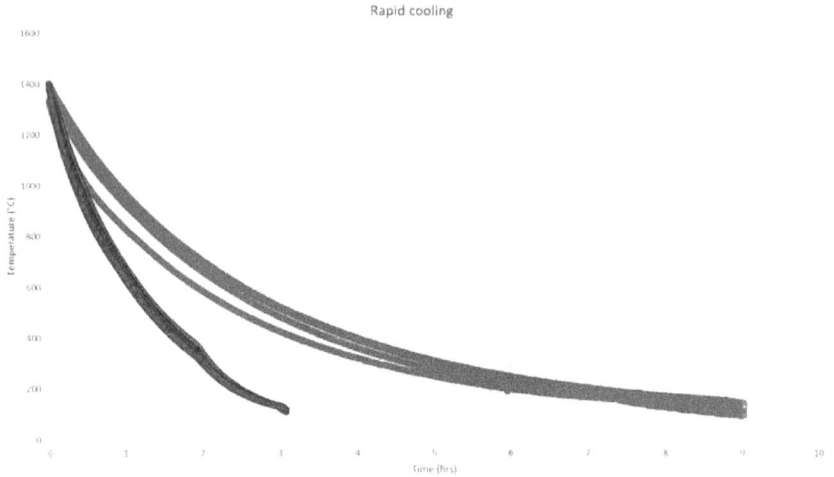

*Figure 4 – Rapid cooling*

*Jet cooling*

Main purpose of the jet cooling is to intensify the gas circulation, mixing the hot gas with cold gas and distributing this mixture over the entire workzone length.

In practical terms, this means that the cold gas entering the bottom of the hot zone in the conventional rapid cooling system is now forced to the top of the hot zone by means of an injector system. Because of its higher density, this cold gas falls down towards the bottom of the hot zone while it absorbs heat from the workload. During this process the density of the gas decreases, the downward movement stagnates and as the density decreases further the gas rises again towards the closed top heat shield and resumes the conventional rapid cooling convection loop. See figure 5 below.

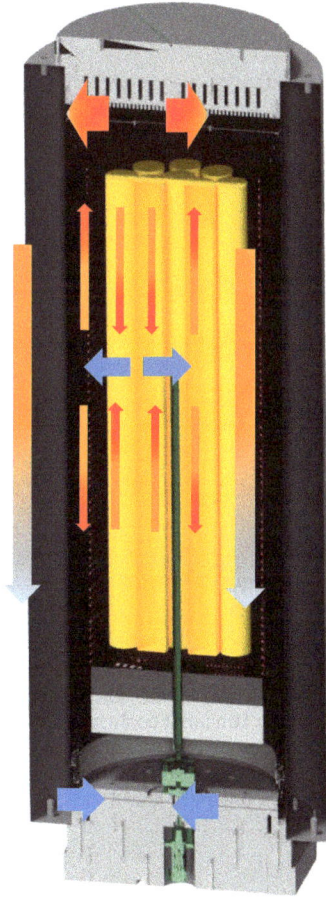

*Figure 5 – Jet cooling schematic*

A lot of research was done in this area, so that we now also get high cooling speeds in a production HIP with solid loads. The cooling rates are of such a degree that they can be used perfectly for metallurgical purposes. See curve in figure 6 below.

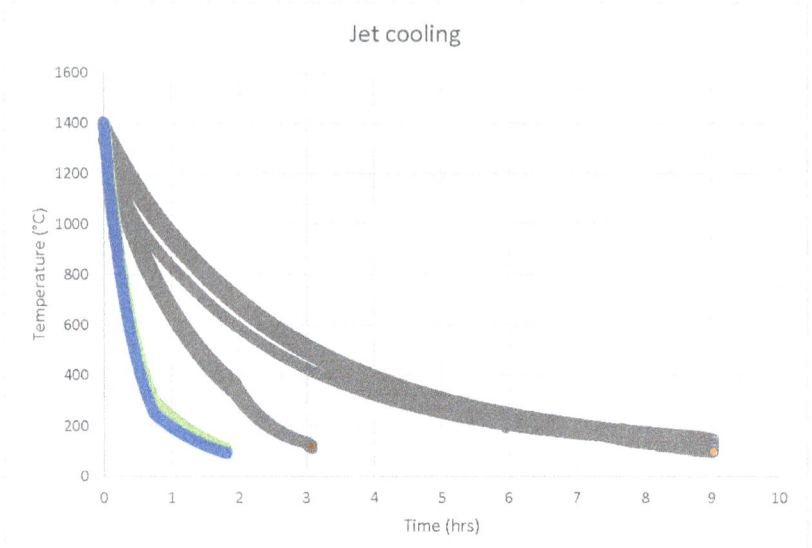

*Figure 6 – Jet cooling*

*Controlled cooling*
Controlled cooling is used to avoid the residual stresses in materials that arise with natural cooling. The controlled cooling makes use of the combination heating and rapid cooling to provide a very accurate curve, even below natural cooling curve.

The controlled cooling is initially slower than the natural cooling, but the desired linear curve eventually crosses the natural cooling curve. At this point, a sophisticated regulation of the rapid cool valve and the heaters holds the cooling rates linear.

See the graph in figure 7.

Materials Research Proceedings **10** (2019) 157-168      doi: http://dx.doi.org/10.21741/9781644900031-22

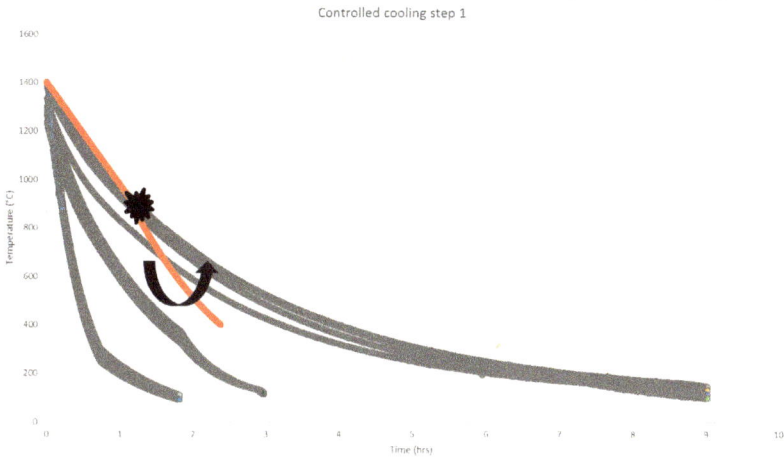

*Figure 7 – Controlled cooling step 1 – until 900 °C*

At a certain moment, the controlled cooling crosses the natural cooling curve (see asterisk in figure 7). At this point, the rapid cooling is activated. The rapid cooling makes the curve bend under the natural cooling curve. Hence the heaters are used to make it linear again. Figure 8 demonstrates this same action at a further point in the curve. This combination of opening the rapid cool valve and using the heaters is used until at the point of decompression.

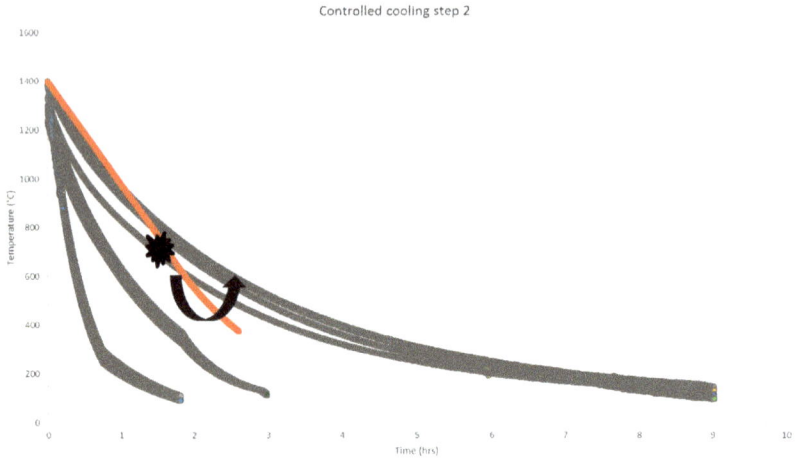

*Figure 8 – Controlled cooling step 2 – until 700 °C*

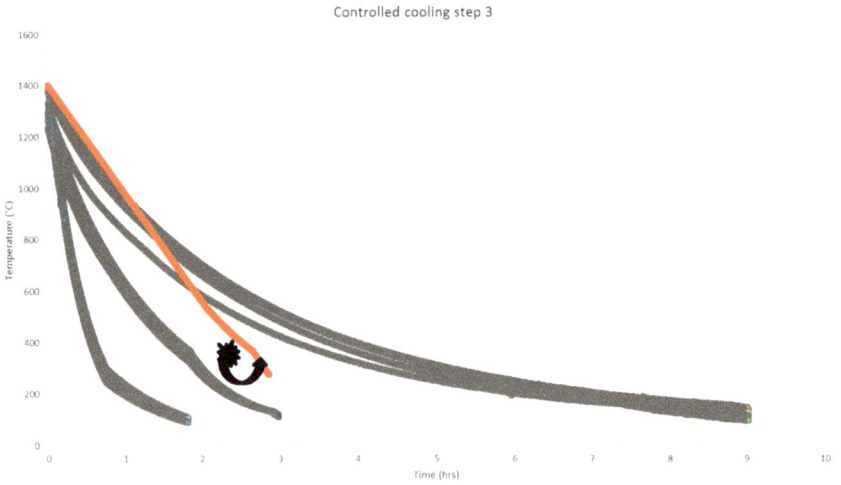

*Figure 9 – Controlled cooling step 3 – until 400 °C*

## Jet cooling acceptance cycle
See figure 10 below for the actual HIP jet cooling acceptance cycle

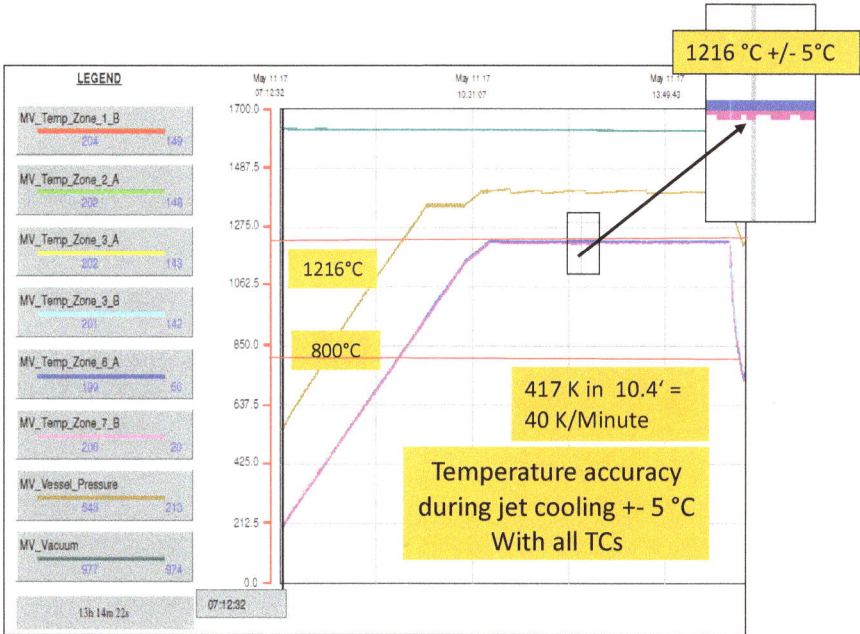

*Figure 10 – Results reached during acceptance cycle*

The prescribed cooling rate of 40 K/minute up to 800°C was exceeded. 46 K/minute was reached. In the range between 1220°C and 850°C, a cooling rate of 51 K/minute was achieved. Below 630°C, rapid cooling was no longer required and was switched off. As indicated on picture 10, temperature accuracies are +/- 5°C better than mentioned in the specification requirements and could be maintained during the whole cycle.
All thermocouples were tested and set in compliance with AMS 2750 E standards.

## Influence of material type and geometry on the results
To stress once again that we obtain good results in real production conditions with real material, we point out that our acceptance cycle took place with ordinary steel. When special alloys are HIPed, such as Ti6AlV4 or AlSi10Mg, even higher speeds become possible. This can be seen from the table 1 below

| Example 1: heat Q required to heat up a load with equal dimensions but consisting of different materials from 20°C up to 1225°C | | | | | | | | | |
|---|---|---|---|---|---|---|---|---|---|
| Material | Density (kg/dm³) | Dimensions (dm) Ø | Length | Weight (kg) | Number | Total Weight (kg) | Specific heat at 20°C (J/(g x K)) | Thermal conductivity at 20°C (W/(m x K)) | Required heat (kJ) |
| S235 | 7.83 | 1.66 | 24.7 | 417 | 6 | 2'500 | 0.47 | 50 | 1407·10³ |
| Ti6Al4V (Grade 5) | 4.45 | 1.66 | 24.7 | 237 | 6 | 1'421 | 0.56 | 7.1 | 938·10³ |
| AISI 10Mg | 2.67 | 1.66 | 24.7 | 142 | 6 | 852 | 0.92 | 119 | 924·10³ |

*Table 1 – Heat Q required to heat up a load*

During the cooling process, the heat must be reduced. In table 2, the values for this process are recorded in the range of 1225°C to 800°C. Compared to steel, the required heat discharge of Ti6Al4V during cooling is substantially smaller.

For this material, a cooling rate increased by approximately 1/3 could be expected.

| Example 1: Heat discharge Q required for cooling down a load from 1250°C up to 800°C with equal dimensions but consisting of different materials | | | | | | | | | |
|---|---|---|---|---|---|---|---|---|---|
| Material | Density (kg/dm³) | Dimensions (dm) (Ø) | Length | Weight (kg) | Number | Total weight | Specific heat at 20°C (J/(g x K)) | Thermal conductivity at 20°C (W/(m x K)) | Required heat (kJ) |
| S235 | 7.83 | 1.66 | 24.7 | 417 | 6 | 2'500 | 0.47 | 50 | 477·10³ |
| Ti6Al4V (Grade 5) | 4.45 | 1.66 | 24.7 | 237 | 6 | 1'421 | 0.56 | 7.1 | 318·10³ |
| AISI 10Mg | 2.67 | 1.66 | 24.7 | 142 | 6 | 852 | 0.92 | 119 | 313·10³ |

*Table 2 – Heat discharge Q required*

For a batch weight of 1421 kg of Ti6Al4V steel, an average cooling rate of 66 K/min. between 1400°C and 800°C can be expected.

The test cycle has been done on material in a cylindrical shape. The higher the ratio of surface / volume, the higher speeds can be expected. E.g. cooling turbine blades will be a lot faster than cylindrical shapes.

**Summary**
With a newly developed jet cooling it is possible to achieve cooling rates of 40°K/minute with large-volume HIP systems (Ø 800 x 2500 mm). The temperature uniformity in the furnace is guaranteed at +/5°K throughout the entire cooling process.

Hot Isostatic Pressing – HIP'17                     Materials Research Forum LLC
Materials Research Proceedings **10** (2019) 169-181     doi: http://dx.doi.org/10.21741/9781644900031-23

# Toughness of Duplex Steel Produced by PM-HIP

Christoph Broeckmann

Institute for Materials Applications in Mechanical Engineering, IWM,
RWTH Aachen University, Augustinerbach 4, 52062 Aachen, Germany

c.broeckmann@iwm.rwth-aachen.de

**Keywords:** Duplex Steel, PM-HIP, Charpy-Toughness, Argon Content

**Abstract.** The most important influencing factors for the toughness of duplex steels are being discussed exemplary at grade AISI 318LN. Focus is given to two major aspects: the embrittlement by σ-phase and the embrittlement caused by residual argon pores. While the formation of σ-phase depends on the cooling rate in the HIP vessel, argon porosity can either be caused by insufficient evaporation prior to HIP or small leakages in the capsule. Toughness is discussed in terms of Charpy tests, taking into account the notch radius as additional parameter. The macroscopic results are reflected by investigations of the microstructure. Toughness of PM-HIP steel is compared to appropriate conventionally produced grades.

## Introduction

Components produced by PM-HIP from corrosion resistent steels with a ferritic-austenitic duplex microstructure are widely used in offshore-applications, in the food industry, in ship building and in the chemical industry [1]. In most of these applications high toughness – particular at low temperatures – is a mandatory requirement. In contrast to austenitic steels, toughness of duplex steel shows a temperature dependent transistion from ductile to brittle behaviour. This is caused by the transition from ductile dimple fracture to brittle transcrystalline cleavage at low temperatures within the ferrite grains. While PM-HIP duplex steels mostly have superior strength and corrosion resistance compared to conventionally produced grades, the toughness issue often leads to discussions.

Factors influencing toughness can be grouped into factors describing the loading and effects of the material's microstructure. While the former contain loading conditions which determine the thermal activation of deformation mechanism, like testing temperature and deformation rate as well as stress triaxiality, the latter factors comprise the crystal lattice, the grain size or non metallic inclusions. In case of duplex steels intermetallic phases strongly determine the toughness of the material. Particular the σ-phase which forms in the temperature range between 940°C and 750°C and the α'-phase which forms at temperatures about 450°C leads to embrittlement of duplex steel even at very small volume fractions [2, 3]. Particular in case of PM-HIP materials, toughness is influenced by two additional effects: the content of oxygen and the content of argon. Stable oxides at the surface of powder particles may not dissolve during HIP and remain as so called prior particle boundaries (PPB's). Bengtsson e. al. [4] revealed by a comparison of two duplex steel powders with different oxygen contents that 1470 ppm of oxygen lead to clearly visible PPB's while no oxides were detectable in hot isostatically pressed powder with 110 ppm. Hämöläinen [5] showed that the oxygen content drastically reduces the Charpy toughness of AISI 318LN duplex steel (see fig. 1).

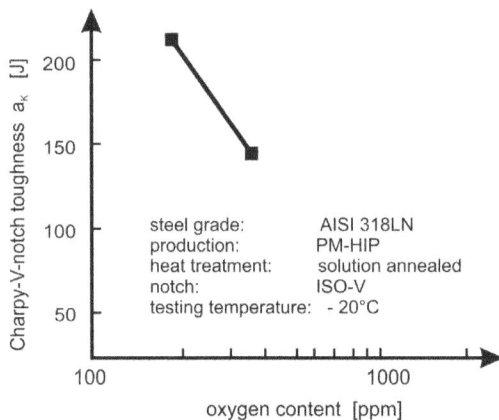

*Figure 1: Influence of oxygen content on Charpy-V-notch toughness, according to [5]*

The second HIP specific factor is Argon. The origin for Ar pores in PM-HIP material can either be contamination of Ar atomized powder or tiny leakages in the capsule. It must be mentioned that using modern production routes for PM-HIP parts, nowadays the argon content usually is below the detection limit. Nevertheless, as small amounts of Ar drastically reduce the toughness, knowledge on the mechanisms that lead to embrittlement by Ar is necessary.

In this study the effects of σ-phase, Ar-content and stress triaxiality has been examined. Moreover, the toughness of duplex steel AISI 318LN is compared to competing production methods like sand casting and open die forging.

**Experimental**

Toughness tests have been performed with duplex steel AISI 318LN (PM2205, ASTM F51, DIN 1.4462, X2CrNiMoN22-5-3) produced by PM-HIP, continuous casting+forging and sand casting. Tab. 1 gives the chemical composition of all three grades tested. The powder for the PM-HIP grade was gas atomized under nitrogen. The oxygen content of the powder was 85 ppm, its apparent density 4.69 g/cm$^3$ and its tap density 5.30 g/cm$^3$. The particle size distribution is given in tab. 2. Two types of capsules have been produced: D110mmxH235mm and D200xH235mm. Capsule material was AISI 304 stainless steel. The big capsules have been consolidated by HIP using a cycle with a holding temperature of 1140°C, a holding time of 200 min and a pressure of 1010 bar. Fig. 2 shows this HIP cycle with the temperature range for σ-phase included.

In order to produce samples with defined Ar-contents 5 "small" capsules have been filled with powder (relative filling density: 66-69%). After leak testing with He, these capsules have been 5 times evacuated down to 10$^{-3}$ mbar and flushed with Ar (1.5 bar). The desired Ar-content was adjusted by the Ar pressure of the final flushing step. These capsules have been consolidated at 1140°C for a holding time of 240 min and a cooling rate of 10 K/min. The pressure was 1000 bar. The final Ar-content was characterized using a gas analyzing device type Extra-Werf with a minimum detection limit of 0.05 ppm. This device is based on gas chromatography.

*Table 1: Chemical composition of the steel grades investigated*

| production route | Chemical composition [weight- %] | | | | | | |
|---|---|---|---|---|---|---|---|
| | C | Si | Mn | Cr | Mo | Ni | N |
| PM-HIP (powder) | 0.015 | 0.68 | 1.02 | 22.20 | 3.09 | 5.19 | 0.170 |
| continuous cast + hot forged | 0.020 | 0.50 | 1.71 | 22.46 | 3.36 | 5.40 | 0.174 |
| sand cast | 0.029 | 0.72 | 1.35 | 23.07 | 2.78 | 6.5 | 0.181 |
| EN 10088 | ≤ 0.03 | ≤ 1.0 | ≤ 2.0 | 20.0 – 23.0 | 2.5 - 3.5 | 4.5 – 6.5 | 0.10 – 0.22 |

*Table 2: Particle size distribution of the gas atomized steel powder*

| particle size [μm] | 500 | 250 | 125 | 106 | 63 | 45 |
|---|---|---|---|---|---|---|
| fraction < [ %] | 100 | 84 | 47 | 37 | 18 | 10 |

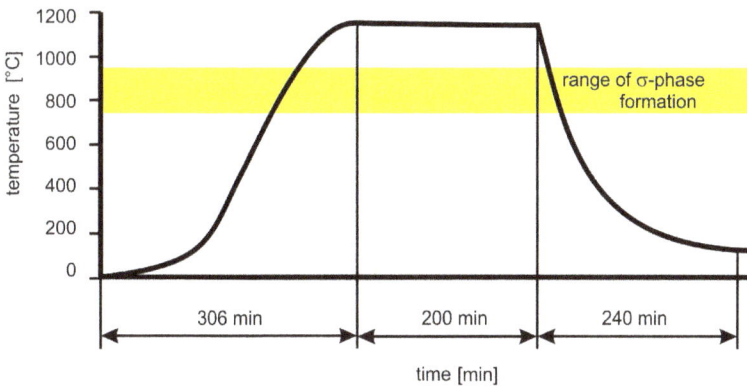

*Figure 2: HIP cycle for the production of samples in an industrial scale plant*

In order to characterize the microstructure light optical sections have been prepared by grinding with SiC paper and polishing with diamond slurry. The microstructure was revealed using Groesbeck-etching. Using this preparation prior to light optical microscopy the austenite appears light brown, the ferrite dark brown and σ-phase blue. The phase constitution was determined using EDS analysis in a SEM Typ Zeiss, Leo 1450VP. SEM was also used to investigate the fracture surfaces. The density of the consolidated samples was measured by He-pyknometry.

Hot Isostatic Pressing – HIP'17                                                          Materials Research Forum LLC
Materials Research Proceedings **10** (2019) 169-181          doi: http://dx.doi.org/10.21741/9781644900031-23

Charpy notch toughness was measured using a pendulum hammer with maximum energy of 300 J. Three samples were tested per temperature, 296 samples were tested in total. The testing temperature was varied between -185°C and +400°C. For this purpose, the specimens were cooled in different liquids or heated in a furnace prior to testing. Liquid nitrogen has been used as cooling liquid for -185°C and a mixture of ethanol and dry ice for -40°C and -75°C, respectively. The temperature in the furnace or the cooling vessel was measured with a thermocouple directly attached to the specimen. After reaching the desired temperature the specimen was placed to the anvil of the hammer and immediately broken within a maximum time period of 3 seconds. Fig. 3 shows the geometry of the Charpy specimens. In order to study the effect of the stress triaxiality the notch radius was varied from 0.25 mm via 1.0 mm to 2.0 mm.

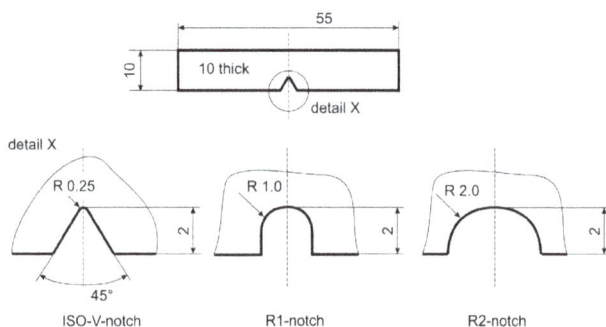

**Figure 3:** *Geometry of Charpy-notch-specimens*

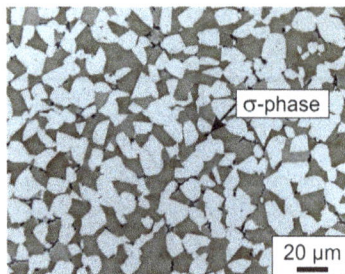

**Figure 4:** *Microstructure of the PM-HIP grade after HIP*

**Results**
The microstructure of PM-HIP 318LN steel in the as HIP condition is shown in Fig. 4. At the boundaries between ferrite and austenite grains the precipitation of σ-phase can be seen as well as at boundaries between ferrite and ferrite grains. In order to remove the σ-phase after HIP most samples were solution annealed at 1060°C for 3 h and subsequently quenched in water. By this

heat treatment the σ-phase could be totally removed. The fractions of all phases were determined by image analysis using a magnification of 500:1 and 1000:1. A number of 3 micrographs corresponding to a total area of 44 mm$^2$ for the as HIP'ed state and 131 mm$^2$ for the annealed state were analyzed. The resulting volume fractions are given in tab. 3 for the as HIP state and for the solution annealed steel.

**Table 3:** *Phase fractions in PM-HIP duplex steel, depending on heat treatment*

| state | phase fraction [vol.- %] | | |
|---|---|---|---|
| | austenite | ferrite | σ-phase |
| as HIP | $50.9 \pm 1.3$ | $46.8 \pm 2.2$ | $1.3 \pm 0.2$ |
| HIP + solution annealed | $50.5 \pm 1.4$ | $49.4 \pm 1.4$ | 0 |

In [6] the composition of each phase has been determined using EDS analysis. The ferrite dissolves about 3.7% more chromium and 2.8% less nickel compared to the austenite. 3% Ni and 8% Mo are dissolved in the σ-phase.

Fig. 5 shows the microstructure of the PM-HIP duplex steel in comparison to material produced by alternative production technologies. Obviously, the forged grade is characterized by a remarkable anisotropy. The individual grains are elongated in the direction of hot forming as seen in the longitudinal section. The transverse section reveals a dispersion of austenite in the ferrite. The microstructure produced by sand casting is isotropic but is determined by a coarse dendritic structure of the austenite.

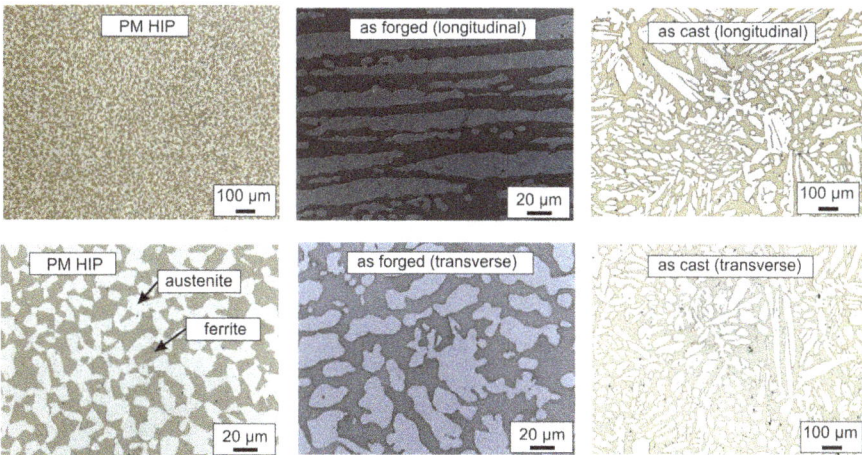

**Figure 5:** *Microstructure of steel AISI 318LN after solution annealing;*
*left: PM-HIP, middle: as forged; right: as cast*

Charpy toughness $a_k$ in this study is plotted against the testing temperature. This allows to discuss the influence of the different effects on toughness in terms of the height of the upper shelf and the shift of the brittle to ductile transition temperature (BDTT). The measured energy to fracture in all diagrams showing toughness results is related to the net cross section of the samples. Fig. 6 shows that the presence of only 1.3 vol.-% of σ-phase in the as HIP state reduces the toughness drastically: Toughness in the upper shelf with 90 $J/cm^2$ is about only one third of the appropriate level of 312 $J/cm^2$ in the precipitation free state. This illustrates that components with thick cross sections made of PM-HIP duplex steel need to be post heat treated prior to service. The local stress triaxiality in the notch root has been varied in fig. 6 by modifying the notch radius. As can be expected an increase of the notch radius leads to a distinctive increase of the upper shelf toughness and a slight shift of BDTT towards lower temperatures. The relevant testing temperature for multiple applications in the offshore industry is -44°C. Although the standard ASTM A988 [7], which gives specifications for mechanical properties for stainless and duplex steel produced by PM-HIP, does not explicitly specifies a minimum toughness value, the toughness plotted in fig. 6 at -44°C should fulfil the requirements of most applications.

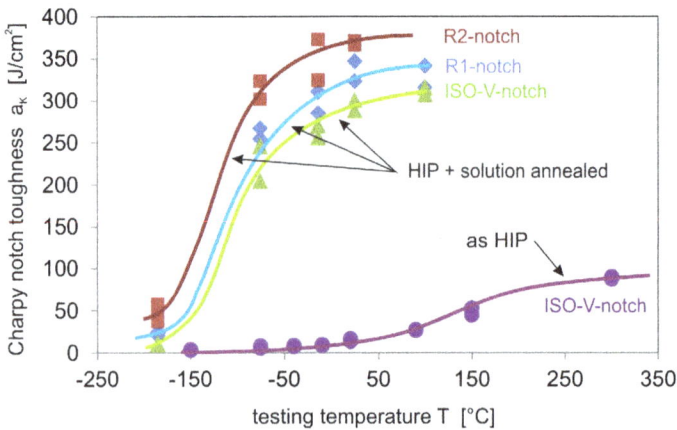

*Figure 6:* *Charpy toughness of PM-HIP duplex steel – influence of heat treatment and notch geometry*

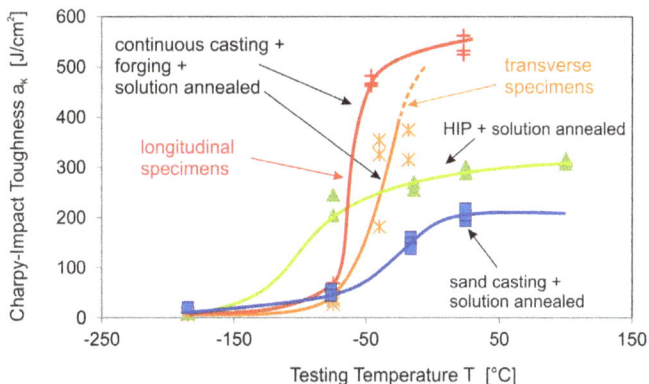

***Figure 7:*** *Charpy toughness of AISI 318LN duplex steel: Comparison of production methods,*
*notch geometry for all specimens: ISO-V-notch*

Fig. 7 shows the influence of the production method on the toughness. The lowest toughness was measured with samples produced by sand casting. Although a level of approx. 200 J/cm$^2$ in the upper shelf represents still a tough material, this value is remarkably lower compared to the PM-HIP grade. Comparing those two routes, the BDTT is lower in the as HIP material and in contrast to the cast grade gives sufficient toughness even at -44°C. The highest toughness is achieved with the forged material. The disadvantage of forged steel is that toughness reduces very rapidly as soon the testing temperature falls below BDTT. This leads to the effect that the PM-HIP steel seems to be superior in applications which run at very low temperatures. As has been investigated in [6] the microscopic appearance of the fracture surface corresponds to these toughness curves: While the PM-HIP grade even in the transition area shows high amounts of ductile dimple failure, cleavage of the ferrite dominates the fracture surface of the forged grade in this temperature range. At very low temperatures only cleavage of ferrite grains is visible at all production routes: Consistently the toughness in the lower shelf does not depend on the production method.

The production of samples with artificial Ar contents lead to voids, filled with Argon, which appear circular in metallographic sections. Fig. 8 shows Ar pores in samples containing 1.0 ppm and 7.1 ppm Ar, respectively. At higher resolution (fig. 9) it can be seen, that at some regions the Ar voids seem to be arranged like a pearl necklace along the former particle boundaries. During HIP crystallographic grains grew across the former particle boundaries proving that the consolidation process performed perfectly.

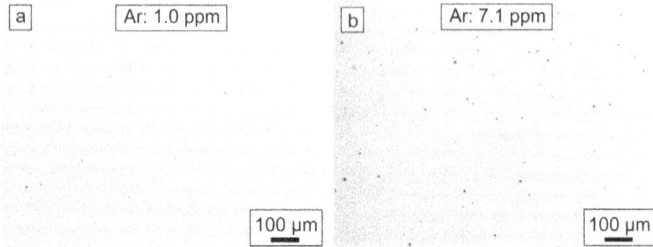

*Figure 8:* Light optical section of HIP samples with different Ar contents

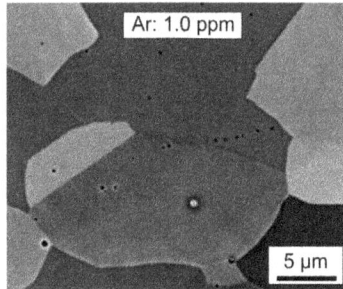

*Figure 9:* SEM section of a HIP sample with 1.0 ppm of Ar

Ar voids drastically reduce the toughness of PM-HIP duplex steel, as figures 10 and 11 point out for samples with two different notch geometries. The green lines represent the condition free of Argon. "< 0.05 ppm" means that the samples contain less Ar compared to the detection limit of the analyzer used. It can be expected that the Ar level is much lower. Already 0.52 ppm reduces the Charpy toughness in the upper shelf and leads to a shift of BDTT towards higher temperatures. 93 ppm of Ar lead to brittle fracture even at higher testing temperature.

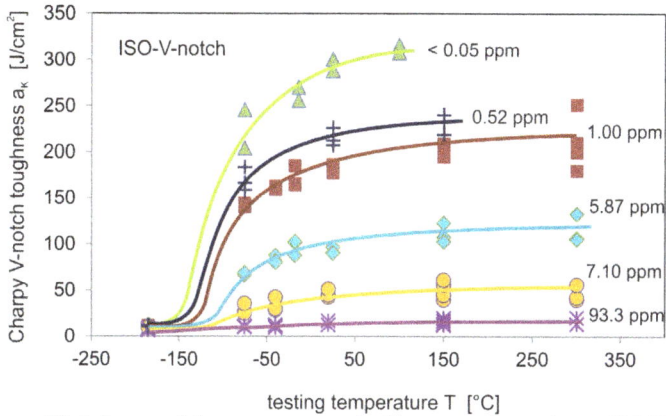

**Figure 10:** *Influence of Argon content on the Charpy notch toughness, ISO-V-notch*

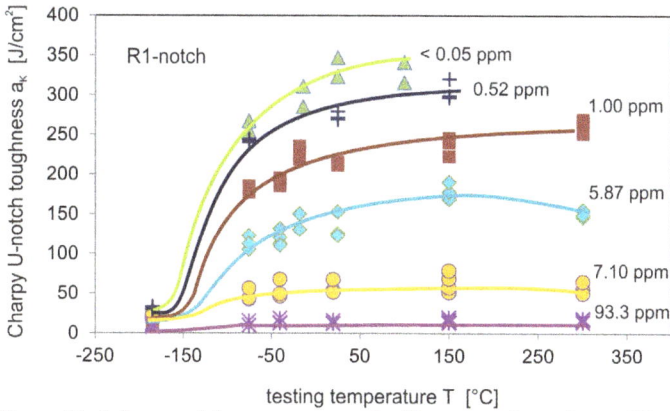

**Figure 11:** *Influence of Argon content on the Charpy notch toughness, R1-notch*

The fracture mechanisms can be studied by looking onto the fracture surfaces of broken Charpy-samples. Fig. 12 shows the fracture surface of a sample with 1.0 ppm Ar, tested at a temperature of 150°C. The high fracture energy indicates a ductile fracture mechanism. This is proven by dimples at the fracture surface, occurring in both phases ferrite and austenite. Individual Ar voids coalesce and form crack like carves (fig. 12 a). Individual Ar pores in the fracture surface no longer appear as spheres, but form a crack like pattern as seen in fig. 12 c. It can be assumed that trapped Ar under high pressure supports this transformation into microcracks during mechanical loading of the material. The same pattern is visible in the sample with 5.8 ppm Ar, tested at 150°C. Although the fracture energy is less compared to the specimen with 1.0 ppm Ar, the microscopic fracture appearance shows still ductile dimples in both phases. Fig. 13 shows microcrack growth including crack tip blunting and local crack opening by plastic deformation. Again it can be assumed that Ar was trapped under high pressure in the pore.

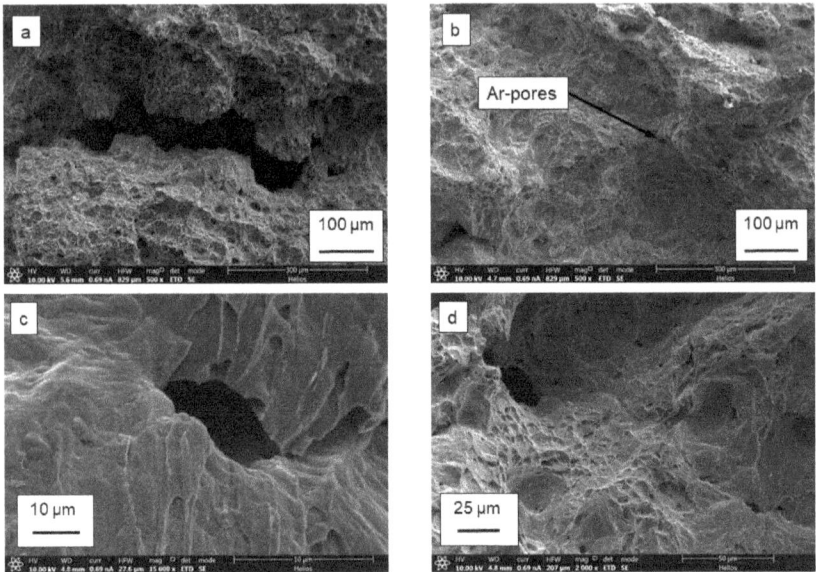

*Figure 12: Fracture Surface (SEM) of a PM-HIP sample with 1.0 ppm Ar, testing temperature $T = 150°C$, $a_k=201 J/cm^2$*

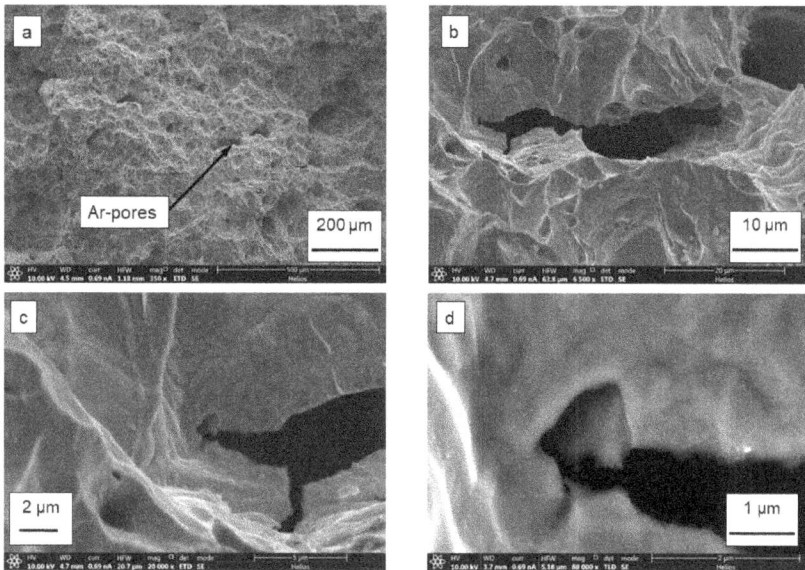

**Figure 13:** *Fracture Surface (SEM) of a PM-HIP sample with 5.8 ppm Ar,*
*testing temperature T = 150°C, $a_k$=91 J/cm$^2$*

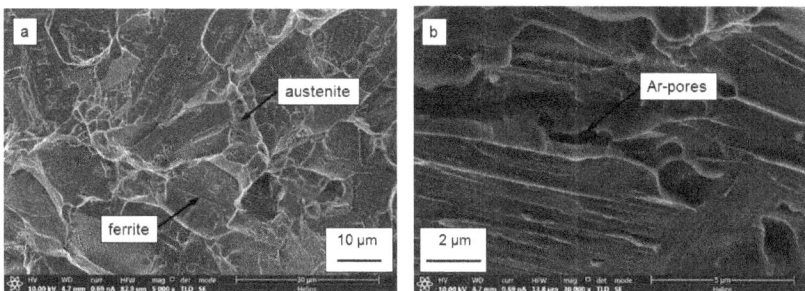

**Figure 14:** *Fracture Surface (SEM) of a PM-HIP sample with 1.0 ppm Ar*
*testing temperature T = -185°C, $a_k$=11 J/cm$^2$*

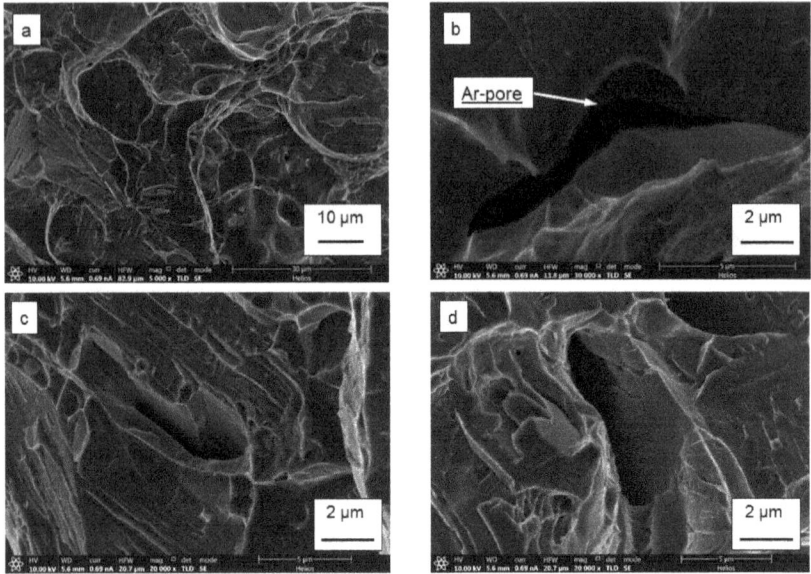

***Figure 15:*** *Fracture Surface (SEM) of a PM-HIP sample with 5.8 ppm Ar,*
*testing temperature $T = -185°C$, $a_k=8$ $J/cm^2$*

At low testing temperature the ferrite fails microscopically brittle by cleavage along crystallographic planes. In case of a sample with 1.0 ppm Ar, tested at -185°C, the biggest part of the fracture surface is covered by those cleavage planes. Some bridging areas through the austenite show local ductile fracture and shallow dimples (see fig. 14). Some Ar voids are visible at the fracture surface (fig. 14b). But in contrast to fracture under conditions leading to upper shelf fracture work, this Ar pore did maintain their shape during fracture and did not behave like a microcrack. It did not develop a sharp crack tip or typical signs of micro crack propagation. Finally, fig. 15 shows the fracture surface of a sample containing 5.8 ppm Ar and being tested at a temperature of -185°C. Again, only small areas of dimples are visible, indicating crack propagation through the austenite. Small lips, typical for cleavage fracture are seen in ferrite grains (see fig. 15 d). One Ar pore shows a crack like appearance (fig. 15 b), but does not reveal any signs of micro crack propagation. It can be concluded that trapped Ar does only slightly influence fracture in the low temperature regime below BDTT. Consistently the Charpy toughness in the lower shelf is only slightly depending on the Ar content.

**Summary**

The effects of manufacuturing method, heat treatment, notch geometry and Ar-content on the Charpy notch-toughness of AISI 318 duplex steel have been investigated. Independent of the testing temperature the toughness of PM-HIP steel is superior to cast material. PM-HIP material shows higher toughness compared to forged steel at temperatures lower than -50°C. Even a very small amount of σ-phase drastically reduces the toughness. A small notch radius leads to an increase of the BDTT and a decrease of the toughness level in the upper shelf. Toughness is very sensitive to the Ar content. Above BDTT Ar voids behave like microcracks which propagate during loading in the Charpy hammer. By this, small amounts of Argon decrease the fracture energy in the upper shelf of the toughness-temperature diagram. 93 ppm of Ar lead to brittle fracture even over the whole temperature range.

**Acknowledgement**
The author would like to thank Mrs. Vanessa Derichs, Mr. Johannes Kunz, Mr. Thomas Güthoff, Mr. Robert Mager, Mr. Henrik Wünsch for their contribution to the study, Mr. Bengt Olof Bengtson from Carpenter Powder Products AB for the delivery of the powder, the companies Deutsche Edelstahlwerke and Schmolz&Bickenbach Guss for delivery of the cast and forged grades and Mrs. Kathrin Horrenkamp from Bodycote HIP GmbH in Haag-Winden, Germany, for performing the Ar-Analysis.

**References**

[1] Charles, J.; Verneau, M.; Bonnefois, B.: Some more about duplex stainless steels and their applications; Stainless Steels Proceedings (1996), pp. 97 – 103

[2] Storz, O.: Einfluss der intermetallischen σ-Phase auf die Gebrauchseigenschaften eines ferritisch- austenitschen Duplex-Stahls; PhD thesis Ruhr-University Bochum, Europäischer Univ.-Verlag; 2007

[3] Nilsson, J.-O.: Overview: Super duplex stainless steels; Materials Science and Technology, (1992) 8, Spp 685-700

[4] Bengtsson, B.-O.; Eklund, A.; Del Corso, G.; Scanlon, J.: Material Properties of PM HIP Stainless Steels; Proceedings of PM2010 World Congress; 10.-14.10.2010, Florenz, pp. 541-545

[5] Hämäläinen, E.; Laitinen, A.; Hänninen, H.; Liimatainen, J.: Mechanical properties of powder metallurgy duplex stainless steels; Materials Science and Technology, 2 (1997) 13, pp. 103-109

[6] Broeckmann, C.; Gütthoff, T.; Wünsch, H.; Bengtsson, B. O. B.; Volker, K.-U.: Zähigkeit von Duplexstahl, hergestellt durch Pulver-HIP; in: Kolaska, H. (Hrsg.): Moderne Fertigungsprozesse – Qualität und Produktivität in der Pulvermetallurgie; Tagungsband zum 29. Hagener Symposium, Hagen, 28.-29.11.2013, pp. 123-146

[7] Standard specification for hot isostatically-pressed stainless steel flanges, fittings, valves and parts for high termperature service, ASTM A988/A 988M-07, 2008

Hot Isostatic Pressing – HIP'17                                            Materials Research Forum LLC
Materials Research Proceedings **10** (2019) 182-189       doi: http://dx.doi.org/10.21741/9781644900031-24

# Precise Prediction of Near-Net-Shape HIP Components through DEM and FEM Modelling

Yuanbin Deng[1,a] *, Anke Kaletsch[1,b], Alexander Bezold[1,c] and Christoph Broeckmann[1,d]

[1] RWTH Aachen University, Institute for Materials Applications in Mechanical Engineering (IWM), Augustinerbach 4, D-52062, Aachen, Germany

[a]y.deng@iwm.rwth-aachen.de, [b]a.kaletsch@iwm.rwth-aachen.de,
[c]a.bezold@iwm.rwth-aachen.de, [d]c.broeckmann@iwm.rwth-aachen.de

**Keywords:** Hot Isostatic Pressing (HIP), Discrete-Element-Method (DEM), Powder Filling, Finite-Element-Method (FEM), Powder Densification, Modelling, Simulation

**Abstract.** In Hot Isostatic Pressing (HIP) of metal powder, anisotropic shrinkage of the capsule induced by inhomogeneity of the initial powder filling density determines the reproducible realization of small geometrical allowances. This becomes a detrimental factor in the manufacturing of near-net-shape components due to their high requirements for the final shape accuracy. This challenge can be solved by precisely predicting and controlling the shrinkage with respect to the filling density via numerical simulation. Using Discrete-Element-Method (DEM), a three-dimensional initial powder density distribution of the whole component is simulated. After being validated by experimental results from metallographic examination, the calculated powder density distribution is mapped to a Finite-Element (FE) model. An in-house developed user defined material (UMAT) Subroutine, which considers both instantaneous plasticity at lower temperatures and rate dependent plasticity at higher temperatures, is utilized for the simulation with ABAQUS. The preliminary experimental validation using lab scale component reveals that the shrinkage induced shape changes during HIP can be accurately predicted by iterative simulations. Furthermore, the influences of local density distribution during HIP are also investigated. In summary, the developed simulation method demonstrates high accuracy in HIP component shape prediction. Therefore, the method can be easily used for designing HIP capsules for large and complex components.

## Introduction

Hot Isostatic Pressing of encapsulated metal powder (PM HIP) is a modern manufacturing process to produce complex and highly specified components using a wide range of metals and alloys. The major advantage of PM HIP is the possibility to produce large and massive near-net-shape (NNS) metal components with complex shapes and excellent mechanical properties. In theory, hot isostatically pressed components should have isotropic shrinkage under isostatic pressure load. However, in practice this is usually not the case. Non-uniform shrinkage of components and even high distortion have been observed. This leads to high cost of post processing and long delivery times.

Theoretically, the three main factors which can lead to the distortion of the final shape of HIPed components are [1–5]: inhomogeneous powder density distribution in the capsule, temperature gradients in the HIP unit, and imperfection of capsule and powder materials. In previous work [6], it has been found that the inhomogeneous initial density distribution of the metal powder in the capsule is the dominating factor that causes the anisotropic shrinkage and distortion of the final component shape (Figure 1).

Hot Isostatic Pressing – HIP'17                                    Materials Research Forum LLC
Materials Research Proceedings **10** (2019) 182-189          doi: http://dx.doi.org/10.21741/9781644900031-24

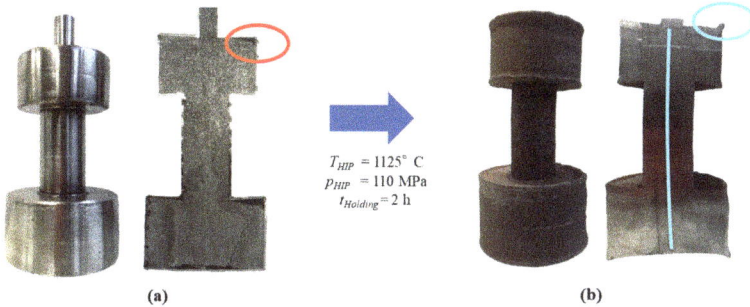

*Figure 1: (a) Welded Capsule (SS304) filled with powder (SS316L, CARPENTER, Melt 500, d=140 µm) prior to HIP, the red cycle shows the zone, which has a poor filling density; (b) Capsule (SS304) filled with powder (SS316L) consolidated to full density after HIP, the blue lines show the bending of capsule and anisotropic deformation*

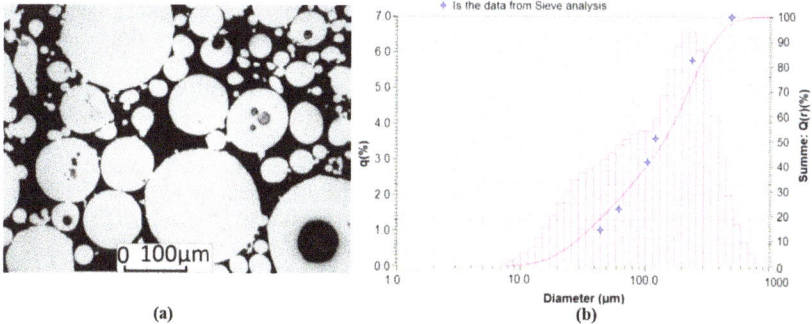

*Figure 2: (a)  A microstructure of powder prior to HIP (Light optical micrograph); (b) Powder particle size distribution (Laser Scattering Particle Size Distribution Analyzer LA 950)*

The powder used in this study has a mean particle size of 140 µm as shown in Figure 2. The results were obtained from dry measurement done by the Laser-Scattering Particle Size Distribution Analyzer Type Horiba LA-950.

This work focuses on the numerical determination of the powder density distribution in the capsule prior to the HIP. Combining the Discrete-Element-Method (DEM) and the Finite-Element-Method (FEM) on the one hand, the powder density distribution is studied numerically. On the other hand, the influences of the local density distribution during HIP are also investigated via simulation. This reduces the need for the often-costly prototypes and trial-and-error processes that are nowadays inherent to capsule design and final shape prediction.

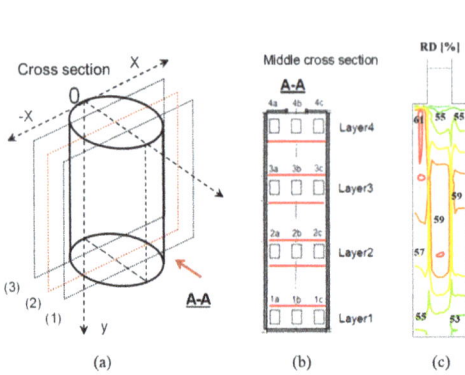

*Figure 3: Arrangement of measured points for Image Analysis; (a) several cross sections were analysed; (b) sketch of measured points (Sub-windows) for the cross section along a radial-axial plane; (c) example of a relative density distribution at the middle cross section of a complex HIPed component. (RD: Relative density)*

## State of the technology

### *"Image Analysis" - Experimental Determination*

Previous investigation [7] showed that the initial powder density distribution in the capsule is strongly influenced by the filling procedure and the pre-compaction process prior to HIP. To analyse the density distribution in the capsule, the powder particles must be "frozen" in their positions after filling and pre-compaction. Therefore, a pressure-less pre-sintering step, which induces very limited overall shrinkage, was applied to build up weak connections between individual particles. After this pressure-less pre-sintering at a temperature close to the HIP temperature for 30 min, the arrangement of the powder particles in the capsule was fixed. In this state, the capsule was cut along several planes (Figure 3(a) and (b)) to obtain samples for the metallographic preparation. Based on the developed "Image Analysis" method, the micrographs of the samples were captured, stored and sorted. The resolution and contrast of all the images were controlled and adjusted to distinguish between porosity and powder particles. The obtained images were digitalized and then converted to the relative density values. These values were then imported into the simulation model (Figure 3 (c)) using an in-house written MATLAB script. However, the determination of the local density distribution is relative time-consuming. Only the relative density on the selected cross planes were measured and the values were extrapolated on the rest areas. The deviation between the determined and real filling density was measured and less than 5%. Due to these limitations, a numerical method was developed to calculate and determine the powder density distribution inside filled capsule.

## Methodology

### *Discrete-Element-Method (DEM) - Numerical Determination*

In recent years, due to the advancement in computer processing speed, numerical simulation of granular flow becomes increasingly effective as an alternative tool to study the behaviours of powder flow, filling and the pre-consolidation process. Since a granular system is composed of individual particles and each particle moves independently of each other, it is difficult to predict the behaviour of a granular system using continuum mechanical models. In this context, the discrete approach developed for particle scale numerical modelling of granular materials has become a powerful and reliable tool. It is considered as an alternate to the continuum approach. This discrete approach is generally referred to as Discrete-Element-Method (DEM).

DEM is a particle method based on Newton's laws of motion. Particles can move with three degrees of freedom (three for translational movement). The particles are defined as soft-particles with consideration of material deformation. During the calculation, a contact model is applied to calculate the contact forces between individual particles. Furthermore, external forces such as gravity or external pressure can be introduced into the simulation as well[8]. All the simulation related parameters are listed in the Table **1**.

*Table 1: Fixed model parameters used in this simulation*

Density $\rho$ 7900 kg/m³
Particle radius $r$ 125 μm
Coefficient of restitution crestitution 0.3
Coefficient of static friction cfriction 0.5
Coefficient of rolling friction crolling 0.5

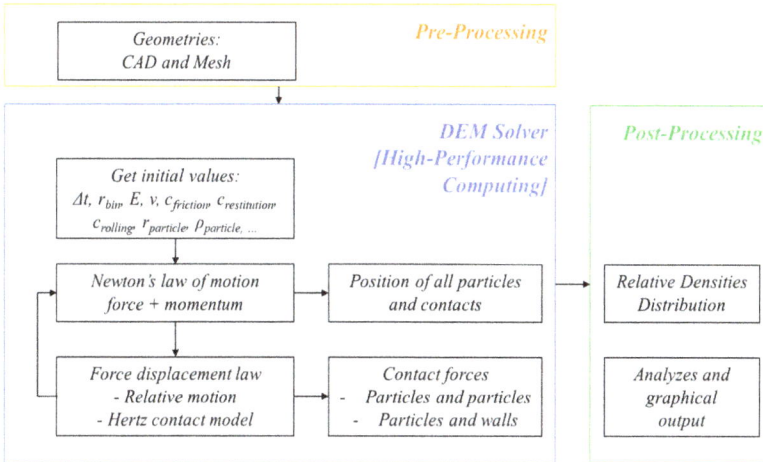

*Figure 4: DEM Modelling Approach for determination of initial density distribution.*

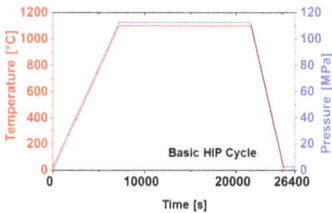

*Figure 5: Temperature and pressure profile for the HIP densification process*

A DEM simulation is divided to three different steps (Figure 4): model design by pre-processing, numerical solution of Newton's equation of motion, export and evaluation by post-processing. After the DEM simulation, the positions of the particles and the contact forces between individual particles as well as between particles and the capsule are included in the output data.

*Coupled process simulation of powder filling via DEM and powder densification via FEM*
The determined relative densities via DEM after the powder filling process were mapped into the FEM model, which was used to simulate the powder densification process. The basic HIP cycle was applied to the densification process (Figure 5). The HIP densification model (capsule material SS304, powder material SS316L) based on FEM was preliminarily implemented in the user- defined material model (UMAT) [9], which is used to simulate the material behaviours in ABAQUS (Figure 6). The initial values, e.g. stress, strain increment, relative densities, were generated and calculated in ABAQUS. Other mechanical properties depending on the temperature were calculated in the UMAT. The time independent plasticity model of Kuhn and Downey [10,11] and the rate dependent plasticity (viscoplasticity) model of Abouaf [12] were coupled in a combined model. After the calculation of the inelastic strain according to this combined model, the stresses and relative densities were updated at the end of each increment. In this way, the calculation continues until the end of the densification process.

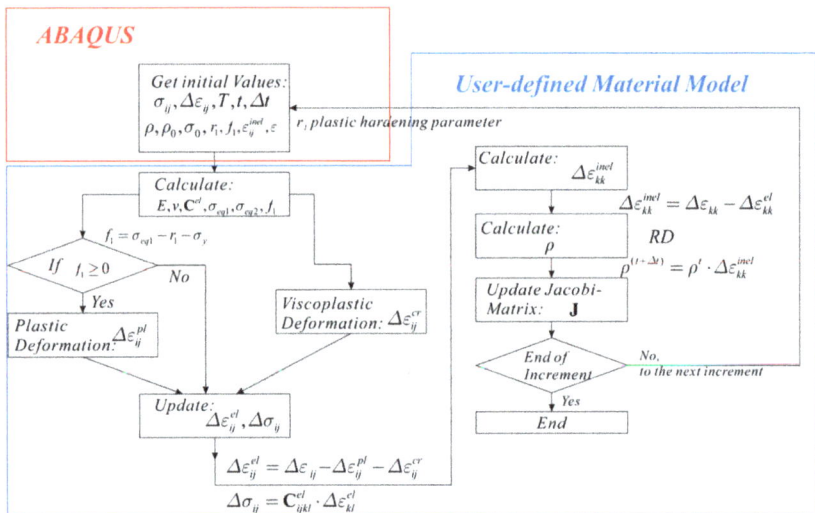

*Figure 6: Flow chart of HIP densification model based on FEM*

**Results and Discussion**
Figure 7 shows a numerical determination of the density distribution in a cylindrical capsule, which was also experimentally determined as shown in Figure 3. With decreased defined powder particle size, the amount of powder particle increases exponentially that leads to the increasing computational time. Before the size of the particle for the simulation completely fixed, the sensitivity analyse was conducted on a small capsule using four discrete particle sizes ($r$ =60 μm, 125 μm, 250 μm, 500 μm). The radius of 125 μm was defined for the simulation, because the simulation using $r$ =125 μm results to a relative small size effect compared to the simulation with size of $r$ =60 μm, meanwhile, the simulation with large particle size can be dramatically reduced. The rheological material properties of SS316L powder relevant for the powder filling simulation

Materials Research Forum LLC
doi: http://dx.doi.org/10.21741/9781644900031-24

were determined and calibrated using a normal flow test with measured angle of repose and a powder flowmeter test with measured filling time.

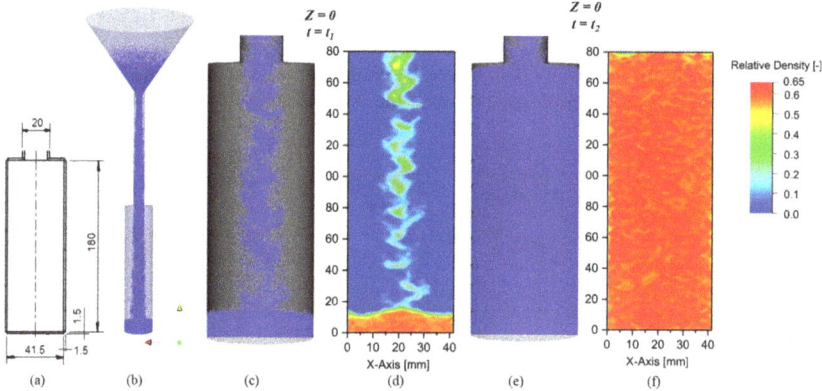

*Figure 7: Filling process using DEM Simulation: (a) capsule geometry, (b) visualization of the filling process, (c) visualization and (d) exported density distribution of the cross section at intermediate step, (e) visualization and (f) exported density distribution of the cross section at the filling end.*

In this case, only mono-sized powder particles with a radius of 125 μm were considered in the simulation. After calibration and validation, the material properties were implemented to simulate the powder filling process in a cylindrical capsule (Figure 7(a)) using a full 3-D DEM model Pre-consolidation processes such as vibration or tapping have not yet been considered. The powder was inserted firstly in a funnel and then flowed into the capsule (Figure 7(b)). The cross-sectional view of the plane Z=0 in intermediate steps during filling are shown in Figure 7(c) and (e). With in-house written scripts, the real-time relative density distributions (Figure 7(d) and (f)) can be obtained. Three-dimensional density distribution was exported to FEM simulation model in which the next processing step is simulated.

The density distributions determined via DEM and "Image Analysis" were mapped in the FEM simulation models (Figure 8 (a) and (b)), respectively. After the capsule filling without any further pre-consolidation process, a high-density region was observed in the middle of the capsule, whereas the positions close to the capsule wall were measured with relatively low filling densities. The difference of the determined densities is ~2% in average. The averaged densities inside the capsule measured by "Image Analysis" and simulated by DEM were 60% and 58%, respectively. For the capsule filling process, only mono-sized powder particles were considered in the simulation. The density distribution might be close to the real case, while the real particle size distribution of the powder material was considered in the simulation.

A script was written and implemented in ABAQUS to map the local densities into the discretized finite element model. Three-dimensional density database was imported into the simulation model prior to HIP, and the densities of the meshes on the cross section is shown in Figure 9, which was determined via DEM simulation.

The FEM models (Figure 8 (a) and (b)) were subsequently used to calculate the HIP densification process. The contours of the final shapes were compared as shown in Figure 8 (c). The contour of the final shape with initial relative density input from DEM simulation is smaller than that with the initial relative density input from "Image Analyse". The largest dimension difference of 3.8% was observed around the top corners of the capsule, possibly due to the difference of the determined density distribution based on two different methods. This deviation can be reduced by improving the prediction with the consideration of particle size distribution via DEM simulation.

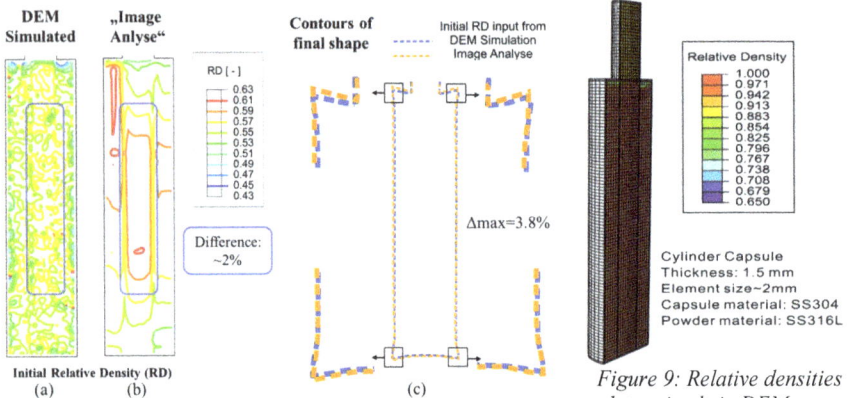

*Figure 8:  Density distribution determined via DEM (a) and "Image Analysis" (b) on cross section of FEM simulation model, (c) Comparison of the contours of final shape.*

*Figure 9: Relative densities determined via DEM were assigned into the simulation model prior to HIP, which was cut on the cross section.*

**Conclusions and Outlook**

In this work, the simulation approach with coupled DEM and FEM modelling has been firstly used to calculate the capsule filling densities and predict the densification behaviour of powder materials during HIP Processes. To predict the powder density distribution inside capsule prior to HIP and the final geometry after HIP precisely, several influencing factors, capsule materials and powder properties, have been considered. The simulation results of the density distribution correspond well with the experimental measurement, which renders the effectiveness of this approach in the precise prediction of the component final shape. To improve the precision of this approach, other factors, such as particle size distribution, different kind of pre-densification processes, will be considered in the following work. Additionally, more complex shaped capsules will also be studied in the future.

**Acknowledgement**

This work was performed under the support from the RWTH Aachen University HPC Compute Project No. rwth0248.

**References**

[1]  H. V. Atkinson, S. Davies, *Metall and Mat Trans A* 2000, *31*, 2981.
https://doi.org/10.1007/s11661-000-0078-2

[2]  E. Olevsky, S. van Dyck, L. Froyen, L. Delaey (Eds.: F. H. Froes, J. Hebeisen, R. Widmer)
1996.

[3]  L. J. R. Trasorras, M. E. Canga, W. B. Eisen, Proc. of the Int. Conf. on Powder Metallurgy
& Particulate Materials 1994.

[4]  T. Shiokawa, Y. Yamamoto, S. Hirayama and Y. Nagamachi, in *Proc. of the Int. Conf. on
Hot Isostatic Pressing* (Ed.: K. Ishizaki) 1992, p. 225.

[5]  M. Abouaf, G. Raisson and E. Wey, in *Proc. of the 3rd Int. Conf. on Isostatic Pressing*
1986, p. 10.

[6]  C. van Nguyen, Numerical Simulation of Hot Isostatic Pressing with Particular
Consideration of Powder Density Distribution and Temperature Gradient, Shaker Verlag 2016.

[7]  C. van Nguyen, A. Bezold, C. Broeckmann, *Powder Metallurgy* 2014, *57*, 295.
https://doi.org/10.1179/1743290114Y.0000000087

[8]  C. Kloss, C. Goniva, Supplemental Proceedings: Materials Fabrication, Properties,
Characterization, and Modeling, Volume 2 2011.

[9]  C. van Nguyen, Y. Deng, A. Bezold, C. Broeckmann, *Computer Methods in Applied
Mechanics and Engineering* 2017, *315*, 302. https://doi.org/10.1016/j.cma.2016.10.033

[10] L. T. Kuhn, R. M. McMeeking, *International Journal of Mechanical Sciences* 1992, *34*, 563.
https://doi.org/10.1016/0020-7403(92)90031-B

[11] H. A. Kuhn, A. Lawley, G. E. Dieter, *Powder metallurgy processing: new techniques and
analyses*, Academic Press 1978.

[12] M. Abouaf, J. L. Chenot, G. Raisson, P. Bauduin, *Int. J. Numer. Meth. Engng.* 1988, *25*,
191. https://doi.org/10.1002/nme.1620250116

Hot Isostatic Pressing – HIP'17                                      Materials Research Forum LLC
Materials Research Proceedings **10** (2019) 190-196        doi: http://dx.doi.org/10.21741/9781644900031-25

# Fabrication of Diamond/SiC Composites using HIP from the Mixtures of Diamond and Si Powders

Ken Hirota [1,a*], Motoyasu Aoki [1,b], Masaki Kato [1,c], Minoru Ueda [2,d],
Yoichi Nakamori [2,e]

[1]Department of Molecular Chemistry & Biochemistry, Faculty of Science & Engineering,
Doshisha University, Kyoto 610-0321, Japan

[2]Metal Technology Co., Ltd, 1-32-2 Honcho, Nakano-ku, Tokyo 164-8721, Japan

[a]khirota@mail.doshisha.ac.jp, [b]ctwb0701@mail4.doshisha.ac.jp, [c]makato@mail.doshisha.ac.jp,
[d]mueda@kinzoku.co.jp, [e]ynakamori@kinzoku.co.jp

**Keywords:** Diamond, SiC, $B_4C$, Liquid-Phase Reaction Sintering, Vickers Hardness

**Abstract.** Diamond/SiC=75/25~50/50vol% composites have been fabricated utilizing a liquid-phase reaction sintering during hot isostatic pressing (HIP) at 1450°C under 196 MPa for 2 h from the mixtures of diamond and Si powders. They were mixed for 30 min in ethanol. After drying, they were compacted uniaxially and isostatically (245 MPa). They were pre-heated at 950°C for 2 h (Process I) or solidified using pulsed electric-current pressure sintering (PECPS) at 1350 or 1450°C for 10 min under 50 MPa in a vacuum (Process II). Both compacts prepared via "Process I" or "II" were densified by Pyrex-glass capsule HIPing (1450°C/2h/196MPa/Ar). The high relative density has been achieved at the composition of diamond/SiC=55/45vol% using Process II. In order to increase Vickers hardness ($H_v$), a small amount of $B_4C$ particles have been added to diamond/SiC= 55/45vol% composites using "Process II" at 1350°C and followed by HIP. The $H_v$ values increased from 37.3 to 40.5 GPa at 5vol% $B_4C$ addition.

## Introduction

As for the continuously increasing need of new novel materials and properties, hard inorganic materials have been attracting a lot of attention. Among them, diamond is the hardest (Vickers hardness $H_v$: ~150 GPa) material with high mechanical properties as shown in Table 1. Since the invention of artificial diamond in 1955 by Bundy *et al.* [1], a lot of works have been done to fabricate polycrystalline diamond (PCD) under the diamond's thermodynamically stable state region [2], such as 5~6 GPa and 1800~1900°C [3]. On the other hand, a few experiments have been attempted in the metastable region [4]. Shimono and Kume [5] studied the fabrication of diamond/SiC composites from diamond and Si powders using a glass-capsule HIP. They reported the bulk density of 3.2~3.3 g·cm$^{-3}$, i.e., the relative density ≥ 90 %, average $H_v$ of 23.5 GPa, and bending strength $\sigma_b$ of ~550 MPa, for the composites fabricated at 1450°C for 30 min at 50 MPa. Recently a new sintering method, *i.e.*, pulsed electric-current pressure sintering, PECPS, has been developed in Japan [6]. This method has some features; it can solidify poor sinterable powders in a shorter soaking time at lower sintering temperature than those of conventional electric furnace [7, 8].

In the present study, we have applied this method to fabricate diamond/SiC composites and compared with those made using the nearly same method [5]. Furthermore, to improve the $H_v$ value, the third hardest $B_4C$ particles have been added to diamond/SiC composites based on thermochemical stability at high temperatures among diamond/SiC matrix. The present paper

Materials Research Forum LLC
doi: http://dx.doi.org/10.21741/9781644900031-25

describes the properties of (PECPS+HIP) sintered diamond/SiC composites and also reports the $H_v$ of $B_4C$ added composites.

## Experimental procedure

*Fabrication of diamond/SiC composites with a small addition of $B_4C$ particles*
As shown in Fig. 1, starting powders were diamond (average particle size $P_s$ of 69 and 3.5 μm, Tomei Diamond, Tokyo, Japan), Si ($P_s$ of 5.0 μm, High Purity Chemical Labo., Saitama, Japan), and $B_4C$ ($P_s$ of 0.5 μm, ibid.). The Si powder was pulverized into $P_s$ of 0.6 μm using a beads-mill, (Easy Nano, AIMEX, Tokyo, Japan) at 1600 rpm for 1.0 h using 2.0 mmφ zirconia balls in ethanol and Ar. The diamond powders of $P_s$ of 69 and 3.5 μm were mixed into a 7:3 ratio [9]. The diamond and pulverized Si powders, furthermore $B_4C$ particles were mixed homogeneously using a vibrating homogenizer in ethanol. After drying, they were mixed with a small amount of 3% acrylic binder and compacted uniaxially and isostatically (245MPa/1min). From our preliminary experimental results, they were pre-heated at 950°C for 2 h (Process I) or solidified using pulsed electric-current pressure sintering (PECPS) at 1350~1450°C for 10 min under 50 MPa in a vacuum (Process II). An acrylic binder is removed by the calcination (950°C/2h/vac.) at Process (I) and PECPS (1350~1450°C/10 min/50MPa) at Process (II). Both compacts prepared via "Process I" or "II" were densified by Pyrex-glass capsule HIPing (1450°C/2h/196MPa/Ar, "O₂ Dr. HIP" Kobe steel, Tokyo, Japan) as shown in Fig. 1, using a liquid-phase reaction sintering [10] between diamond and Si at higher temperatures than the melting point of Si (1414°C) by glass-capsule HIPing (1450°C/2 h/196 MPa/Ar).

Table 1 Physical and mechanical properties of diamond, silicon carbide β-SiC and B₄C

| Property | symbol | unit | diamond cubic | β-SiC cubic | B₄C hexa. |
|---|---|---|---|---|---|
| Theoretical density | $D_s$ | Mg·m⁻³ | 3.513 | 3.21 | 2.515 |
| Vickers hardness | $H_v$ | GPa | ~150 | 28 | 33.5 |
| Melting temperature | $T_m$ | °C | 2450 | 2730 | 2450 |
| Fracture toughness | $K_{IC}$ | MPa·m¹ᐟ² | 2.0 | 4.0 | 3.1 |
| Young's modulus | $E$ | GPa | 1050 | 430 | 385 |
| Thermal expansion coefficient | $\alpha$ | 10⁻⁶·K⁻¹ | 0.8~1.2 | 4.5 | 2.3 |
| Thermal conductivity | $\sigma_T$ | W·(mK)⁻¹ | 600~2000 | 270 | 37 |

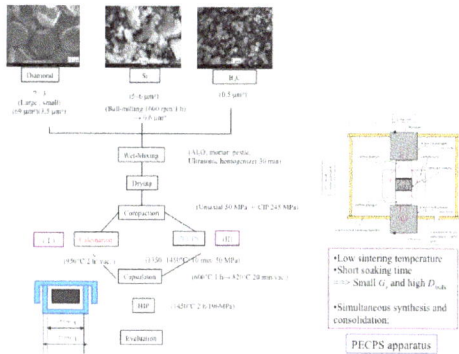

*Fig. 1 Flow chart for the fabrication of diamond/SiC composites.*

*Characterization of samples*
Crystalline phases were identified by X-ray diffraction (XRD) analysis. Bulk densities of sintered samples were measured by Archimedes method. Microstructural observation on the fracture surfaces of sintered samples was performed with a field-emission type scanning electron

Hot Isostatic Pressing – HIP'17                                    Materials Research Forum LLC
Materials Research Proceedings **10** (2019) 190-196      doi: http://dx.doi.org/10.21741/9781644900031-25

microscope and their average grain sizes $G_s$ were determined by an intercept method [11]. Then, $H_v$ was evaluated with an applying load of 9.8 N and a duration time of 15 s with a Vickers hardness tester.

**Results and discussion**

*Difference between Process I and II*

Fig. 2 displays the XRD patterns of diamond/SiC composites with various SiC contents from 35 to 50 vol% or 35 to 100 vol%, fabricated as follows; (a)~(d), calcined (950°C/2h/vac.) via "Process I", and (e)~(h) sintered by PECPS (1450°C/10min/50MPa/vac.) via "Process II", then followed by HIP (1450°C/2h/196MPa/Ar). Big difference between "Process I" and "II" is; i) with increasing the intended content of SiC from 35 to 50 vol%, amount of graphite 2H-C in the HIPed composites decreased in (a) to (d) via "I", on the contrary the amount of graphite increased in (e) and (f) via "II", and ii) the remained content of Si in the composites were still high in XRD patterns of (b)~(d) via "I", however, in "II" the amount of unreacted Si is not so high without the diamond/SiC=30/70 vol% composition, suggesting that when a small amount of Si was added to the diamond powders, and PECPSed at 1450°C unreacted diamond might transform into graphite. This might be explained by that the dense microstructures of diamond/SiC after PECPS at 1450°C might block a direct-contact between diamond and remained liquid Si and resulted in suppression of their liquid state reaction during HIPing. Here, the diffraction peaks of SiC phase in (h) shifted to the lower angles a little than those for pure SiC might be explained by the bulk sample's inclination to the XRD measuring datum surface. Fig. 3 (i) to (iv) display the relative densities $D_r$, *i.e.*, $D_r=D_{obs}/D_x$, here, $D_{obs}$ and $D_x$ are bulk and theoretical densities, respectively, of diamond/SiC composites fabricated by two processes, i.e., via "I" and "II" before and after HIPing denoted as (i), (ii) and (iii), (iv), respectively, as a function of intended content of SiC. In the $D_r$ before HIPing the values of (i) only calcined materials are much lower than those of PECPSed (iii), it might be due to activated pressing during heat treatment.

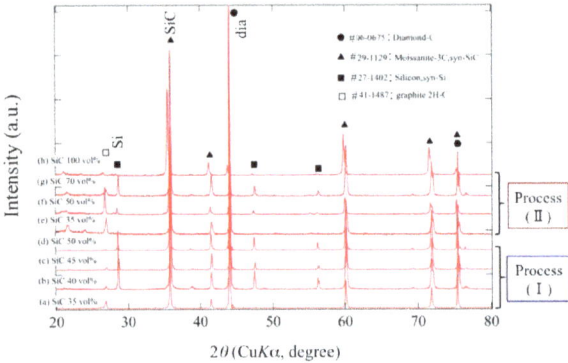

*Fig. 2 XRD patterns of diamond/SiC composites: (a)~(d), calcined (950°C/2 h/vac.) and (e)~(h) sintered by PECPS (1450°C/10min/50MPa/ vac.), followed by HIP (1450°C/2h/196MPa/Ar).*

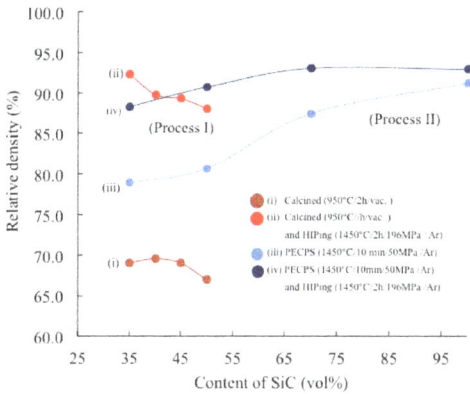

Fig. 3 Relative densities of diamond/SiC composites fabricated under various conditions as a function of SiC content.

However, after HIPing the difference in the $D_r$ between "I" and "II" as shown in (ii) and (iv) is a little. Here, it should be noted that in "I", there is a decline tendency of $D_r$ for both (i) and (ii), on the contrary, an accession tendency of $D_r$ for both (iii) and (iv) in the SiC rich region.

Then their microstructures were observed. Fig. 4 shows SEM photographs of the fracture surfaces of diamond/SiC composites with 65/35 and 50/50 vol% compositions. The upper photographs (a) and (b) are the composites fabricated via "I", and the lower (c) and (d) via "II", both followed by HIPing. Large diamond particles and small (gray: SiC and black: diamond) grains are observed, and their relative densities $D_r$ are also shown to the right side of photograph. In "I", a high $D_r$ (93.3%) is achieved in diamond/SiC=65/35vol%, and in "II" a high $D_r$ (90.7%) is achieved in diamond/SiC=50/50vol%. It is clear that dense grain-boundary consisting of fine SiC and small diamond grains could bring a high $D_r$, in addition to this, good contact, or good adhesiveness between large diamond particle and grain-boundary also has much effect on the densification. Based on these data, in the SiC rich composites, "II" was expected to show higher $D_r$. Fig. 5 shows XRD patterns of diamond/SiC composites added with $B_4C$, their content changed from 0, 1, 3 and 5 vol%: (a)~(d) materials sintered by PECPS at 1350°C and (e)~(f) sintered by additional HIP (1450°C/2h/196MPa/Ar). With increasing the amount of $B_4C$ addition after PECPS, a lot of Si and $B_4C$ diffraction peaks were observed around $2\theta=26.5°$ and $2\theta=37.5°$, respectively.

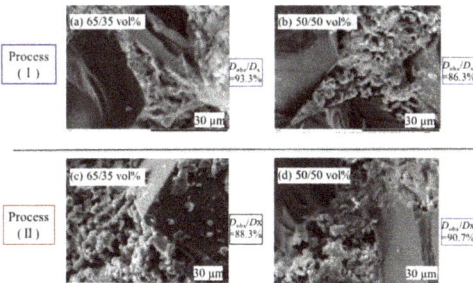

Fig. 4 SEM photographs for the fracture surfaces of diamond/SiC composites: (a)~(b), calcined (950°C/ 2h/vac.) and (c)~(d) sintered by PECPS (1450°C/ 10min/196MPa/Ar), followed by HIPing (1450°C/ 2h/196MPa/Ar).

Hot Isostatic Pressing – HIP'17                                    Materials Research Forum LLC
Materials Research Proceedings **10** (2019) 190-196          doi: http://dx.doi.org/10.21741/9781644900031-25

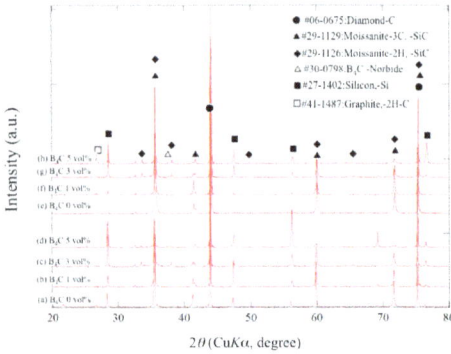

*Fig. 5 XRD patterns of diamond/SiC=55/45vol% composites added with $B_4C$: (a)~(d) sintered by PECPS at 1350°C and (e)~(f) sintered by HIP (1450°C/2h /196MPa /Ar).*

Furthermore, it should be noted that the $B_4C$ addition enhanced the formation of both cubic (#29-1129) and hexagonal (#29-1126) SiC at the same time at 1350°C PECPS.After HIP at 1450°C, $B_4C$, SiC and graphite were recognized in the composites up to 5 vol% addition. The values of $D_{obs}$ and $D_r$ of composites were measured and calculated, respectively.

Fig. 6 represents the $D_r$ for composites as a function of $B_4C$ content. They were decreased once at 1.0 vol%, and then gradually increased until about 92.4% with 5.0vol% addition. Fig. 7 shows their microstructures observed by SEM using back-scattered electron image (BEI); (a) 0, (b) 1.0, (c) 3.0 and (d) 5.0 vol% $B_4C$ added composites; as BEI proves that the heavier elements give brighter image than those of the lighter elements, therefore, both large dark gray blocks and small particles in the grain-boundaries, are large and small diamonds, respectively, and bright grain-boundary matrix might be SiC. However, in these photographs $B_4C$ particles could not be recognized or distinguished from small diamonds among gain-boundaries due to both fine particles and as the nearly same light elements as diamond.

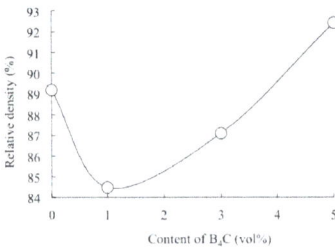

*Fig. 6 Relative densities of $B_4C$-added diamond/SiC =55/45vol% composites as a function of $B_4C$ content fabricated using PECPS at 1350°C for 10 min under 50 MPa in a vacuum, followed by HIP (1450°C/2h/196 MPa/Ar).*

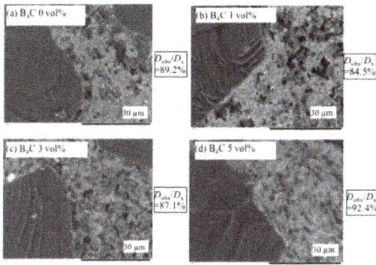

*Fig. 7 SEM (back-scattered image:BEI) photograph of for the fracture surfaces of B₄C-added diamond/SiC=55/45 vol% composites fabricated using PECPS at 1350°C for 10 min under 50 MPa in a vacuum, followed by HIP (1450°C/2h/196MPa/Ar).*

Then, their mechanical property, especially, hardness was measured. Vickers hardness $H_v$ of composites was shown in Fig. 8 [I]. Though much scattering is observed with increasing $B_4C$ content, the average value of $H_v$ increased gradually from around 37.3 to 40.5 GPa at 5 vol% $B_4C$ addition. In Fig. 8 [II], an laser microscopic image of 'Vickers indent", a SEM (BEI) and an energy-dispersive spectroscopic (EDS) images of the microstructure of 5.0 vol% $B_4C$ added composite are presented. The indent size is very small which proves high hardness, and by comparing SEI and EDS images, it is clear that each particle, such as small diamond, $B_4C$ and SiC are homogeneously dispersed. This homogeneous dispersion, thermal expansion coefficient $\alpha$ difference between SiC ($\alpha$:4.3×10⁻⁶ K⁻¹) and $B_4C$ ($\alpha$:2.3×10⁻⁶ K⁻¹) and high Young's modulus (385~430 GPa) induced high compression stress into SiC grain-boundary, which might bring higher $H_v$ value than those (about 23.5 GPa) reported previously [5]

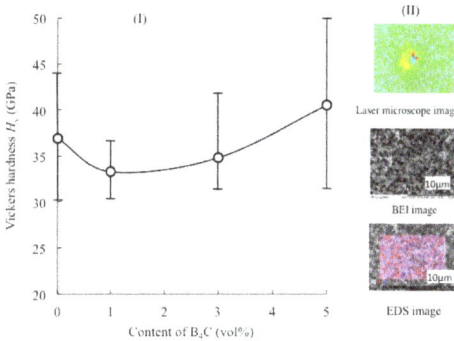

*Fig. 8 (I); Vickers hardness $H_v$ of diamond/SiC=55/45vol% composites sintered by both PECPS & HIP as a function of $B_4C$ content, (II); microphotographs of Vickers indents, upper: laser microscope, middle: BEI figure, lower: EDS image.*

## Conclusion

Dense diamond/SiC=50/50vol% composites have been fabricated from the mixtures of diamond and Si powders using pulsed electric-current pressure sintering (PECPS) at 1350°C for 10 min under 50 MPa in a vacuum, followed by HIP at 1450°C for 2 h under 196 MPa in Ar with a liquid state sintering. This composite reveals high Vickers hardness $H_v$ of 37.3 GPa. Furthermore, a small amount of $B_4C$ addition enhanced the $H_v$ up to about 40.5 GPa due to high compression stress induced into SiC boundary matrix.

**Acknowledgement**

This work was financially supported by the Program for the Strategic Research Foundation at Private Universities, 2013-2017, the Ministry of Education, Culture, Sports, Science and Technology, Japan (MEXT). The authors thank to Dr. S. Hosomi of Tomei Diamond Co., Ltd., for supplying various kinds of diamond powders. Also the authors thank to Dr. Taguchi of Doshisha University for measuring XRD of diamond/SiC composites and advice about their data.

**References**

[1] F.P. Bundy, H.T. Hall, H.M. Strong, R.H. Wentorf, Jr., Man-Made Diamonds, Nature, 176 (1955) 51-55. https://doi.org/10.1038/176051a0

[2] R. Berman, F. Simon, On the Graphite-Diamond Equilibrium, Zeit. Electrochem., 59, 5 (1950) 333-38.

[3] H. Sumiya, N. Toda, S. Satoh, Growth rate of high-quality large diamond crystals, J. Crystal Growth, 237-239 (2002) 1281-1285. https://doi.org/10.1016/S0022-0248(01)02145-5

[4] S. Kume, K. Suzuki, H. Yoshida, Hot Isostatic Pressing of Diamond-Containing Inorganic Composites, pp. 53-55 in Physical Properties of Composites, Proceedings o 125th TMS Annual Meetings Symposium, The Minerals, Metals and Materials Society, Warrendale, PA, 1996

[5] M. Shimono, S. Kume, HIP-sintered composites of C(diamond)/SiC, J. Am. Ceram. Soc., 87, 4 (2004) 752-55. https://doi.org/10.1111/j.1551-2916.2004.00752.x

[6] M. Tokita, Trend in Advanced SPS Spark Plasma System and Technology, J. Soc. Powder Tech. Jpn., 30 (1993) 790-804. https://doi.org/10.4164/sptj.30.11_790

[7] K. Hirota, Y. Takaura, M. Kato, Y. Miyamoto, Fabrication of Carbon Nanofiber(CNF)-Dispersed $Al_2O_3$ Composites by Pulsed Electric-Current Pressure Sintering and their Mechanical and Electrical Properties, J. Mater. Sci., 42[13] (2007) 4792-4800. https://doi.org/10.1007/s10853-006-0830-0

[8] K. Hirota, M. Shima, X. Chen, N. Goto, M. Kato, T. Nishimura, Fabrication of dense $B_4C$/CNF composites having extraordinary high strength and toughness at elevated temperatures, Mater. Sci. Eng. A, 628 (2015) 41-49. https://doi.org/10.1016/j.msea.2015.01.020

[9] S. Yerazunis, J.W. Bartlett, A.H. NIssan, Packing of Binary Mixtures of Spheres and Irregular Particles, Nature, 195 [7] (1962) 33–35. https://doi.org/10.1038/195033a0

[10] J.E. Marion, C.H. Hsueh, A.G. Evans, Liquid-Phase Sintering of Ceramics, J. Am. Ceram. Soc., 70 [10] (1987) 708–713. https://doi.org/10.1111/j.1151-2916.1987.tb04868.x

[11] M.I. Mendelson, Average Grain Size in Polycrystalline Ceramics, J. Am. Ceram. Soc., 52, (1969) 443–446 [6] (2009) 607-616.

Hot Isostatic Pressing – HIP'17    Materials Research Forum LLC
Materials Research Proceedings **10** (2019) 197-202    doi: http://dx.doi.org/10.21741/9781644900031-26

# TeraPi – A 3.5-Meter Diameter Hot Zone Unit Enables HIPing of Large Components

Dr. Anders Eklund[1,a*], Dr. Magnus Ahlfors[2,b]

[1]Quintus Technologies AB, Quintusvägen 2, SE-72166 Västerås, Sweden.

[2]Quintus Technologies, LLC, 8270 Green Meadows Drive North, Lewis Center, OH 43035, USA

[a]anders.eklund@quintusteam.com, [b]magnus.ahlfors@quintusteam.com

**Keywords:** Tera-HIP, Large Components, Powder Metallurgy, Near-Net-Shape (NNS), Forgings, Energy, Volume Production

**Abstract.** Hot isostatic pressing (HIP) has been known for more than 50 years, and is considered today as being a standard production route for many applications. The HIP process applies high pressure (50-200 MPa) and high temperature (300-2,500°C) to the exterior surface of parts via an inert gas (e.g., argon or nitrogen). The elevated temperature and pressure cause sub-surface voids to be eliminated through a combination of mechanical deformation, plastic flow and diffusion.

The largest HIP unit operated in the world today has a hot zone diameter of 2.05 meters which is very big. However, there are even bigger components produced that would benefit from a HIP treatment, but which cannot be HIPed today because of the size. These components could be pump house and valve castings for nuclear power plants, forged parts for reactors or components for aerospace engines for example.

This presentation will cover which types of components and markets that can benefit from this size of HIP. It will also be explained how it is possible to operate an extra-extra-large HIP like the TeraPi and the technical concept together with performance details.

## Motivation

Quintus Technologies AB (at that time Avure Technologies) started a study in 2010 named project Novel HIP. The idea was to expand and brake the boundaries of how big a HIP can be. The question was, can a 10-meter diameter HIP be manufactured? The project was sponsored by Vinnova, the Swedish Organization for innovations and research.

New concepts and designs were invented and patented, i.e. the assembly of the frame, see Fig. 1. It was a unique light weight frame solution for easy transportation, winding at site, assembly at site and very important to have forging sizes available from the suppliers. The total weight of the frame for a diameter of 10 meter and height 8 meter is 8000 ton.

Another concept change was to develop a new pressure vessel cylinder for hot zone diameters over 2 meters. Due to the large load weights involved, 50-75 tons and the weight of the pressure vessel cylinder itself of about 300 tons, the decision was made to move the cylinders instead of the frames. Earlier concept design of HIP units from Quintus has the frames moving. Due to the fact that the frames will grow more in weight than the vessel cylinders, Tera-HIP units now has the frames positioned in a pit, and the HIP pressure vessel will be moving. The maximum operating pressure (MOP) was targeted to be 1050 Bar. See Fig. 2.

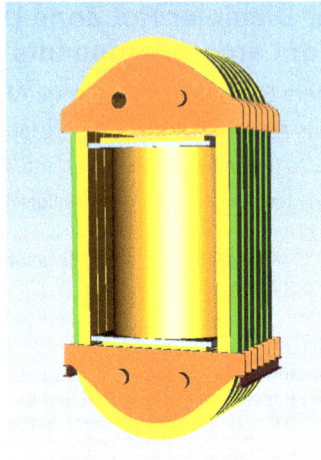

*Fig. 1. Frame for pressure vessels up to 10 meters in diameter and 8 meters in height and MOP of 1000 Bar.*

Also, a new furnace concept had to be developed to effectively HIP and heat treat large parts in less than 24 hours. See Fig. 3. The maximum operating temperature for the furnace was targeted to 1250 °C. That would cover most steel, stainless steel, Titanium, Inconel and Aluminum details predicted to be HIPed. The material of the furnace would be Molybdenum, since it shows great durability and form stability over long periods of usage.

To sum up, the general design requirements and goals were,
- ASME design
- Designed for greater than 98%+ uptime
- Designed for half of today's installation time
- Manufacturing time not to exceed 24 months
- Maximum temperature deviation of +/- 10C during heating/cooling
- Minimum calculated fatigue life of 10,000 cycles…30+ years
- Total load weight of about 75 tons
- Cycle time not to exceed 24 hours
- Conventional technology!!

The outcome of the project would be the QIH TeraPi. A Tera-HIP unit with a hot zone of 3.14 meters, a hot zone height of 5.0 meters and MOP of 1050 Bar, maximum operating temperature of 1250 C, and a variable pay load of 50-75 tons. Even though many new concepts were introduced and patented and tested the TeraPi HIP units were going to use conventional technology. No new concepts were to be used for the Novel HIP.

Hot Isostatic Pressing – HIP'17                                    Materials Research Forum LLC
Materials Research Proceedings **10** (2019) 197-202          doi: http://dx.doi.org/10.21741/9781644900031-26

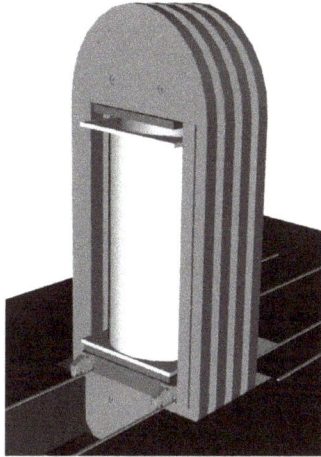

*Fig. 2. Moving cylinder concept on a rail carriage, fitted with top- or bottom loading.*

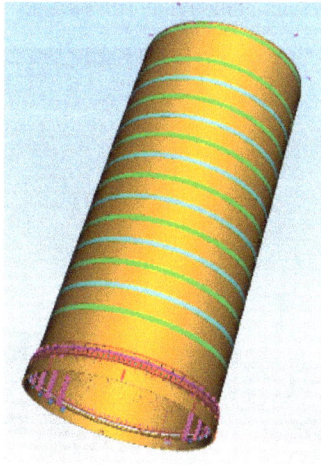

*Fig. 3. Furnace concept for large Tera-HIP units with hot zone diameters over 2 meters in diameter.*

199

Hot Isostatic Pressing – HIP'17                                     Materials Research Forum LLC
Materials Research Proceedings **10** (2019) 197-202          doi: http://dx.doi.org/10.21741/9781644900031-26

**Game changers**

The major achievements in the project Novel HIP can be summarized as below,

- Large diameters, over 3.0 meters (3.14 m = 123" ID)
- Running cost below 1.00 SEK/kg
- Possible to modularize for easy manufacturing and installation
    - No overhead crane
    - No pit
- High degree of automation for easy handling of the loads
- New frame construction with multiple frames
- New furnace construction for handling large volumes of hot gas
- The solutions make the foundation easier
- The solutions minimize Argon risk
- The solutions have noise protected pump unit
- Separate floors for operation and maintenance
- ASME approval

In Fig. 4, one can see a layout of the TeraPi HIP unit with the introduced developments mentioned earlier. All process unit operations, i.e. furnace, gas compressors, water cooling, power supply, VRT's, etc. are all proven technologies and the components have been tested and have been used for many years. One feature which can be seen in the Fig. 4, is that there are separate floors for maintenance and operations. Two load stations are also included for maximum productivity and efficiency of the TeraPi HIP system.

*Fig. 4. The Quintus QIH TeraPi HIP system.*

Other features are an automated handling system of the load and furnace to minimize operator handling errors.

Hot Isostatic Pressing – HIP'17                                    Materials Research Forum LLC
Materials Research Proceedings **10** (2019) 197-202       doi: http://dx.doi.org/10.21741/9781644900031-26

Other sizes of the Tera-HIP systems have been designed after discussions with potential customers world-wide. Most notable are the QIH-T140, with a hot zone diameter of 3.55 meter and height 4.0 meter. And the QIH-Tera4me, with a hot zone diameter of 4.0 meters and height 4.0 meters. These concept designs have been modified so bottom loading is possible. This due to the large weights of the bottom closure plate and the heavy loads that will go in to these HIPs. An example of the bottom loaded Tera-HIP system can be seen in Fig.

*Fig. 5. Bottom loaded Tera-HIP unit for even larger hot zone sizes.*

**Applications**

The Tera-HIP concept has an advantage when it comes to handle very large components for the Energy sector, Aerospace or Defense. Regardless if one manufactures the parts through Powder Metallurgy-Near Net Shape or by making large Castings/Forgings, the Tera-HIP systems can handle most components today that have long lead-times. The lead-times for large forgings can be years instead of months, and that is where the PM/HIP-NNS route has a huge advantage. Examples of some forged parts that instead could be produce by HIP can be seen in Fig. 6.

*Fig. 6. Examples of forged parts for Reactor Pressure Vessels. (Left) Core region shell, (Center) Bottom petal, (Right) Closure head.*

By using Powder metallurgy-HIP for Near-Net Shape (NNS) other advantages can be found. For example, alloy systems that are difficult to produce in the traditional AOD-casting route like XM13, 6-moly stainless steels, newly developed Hyper-Duplex for offshore to minimize corrosion and not yet invented alloys. An example can be seen in Fig. 7.

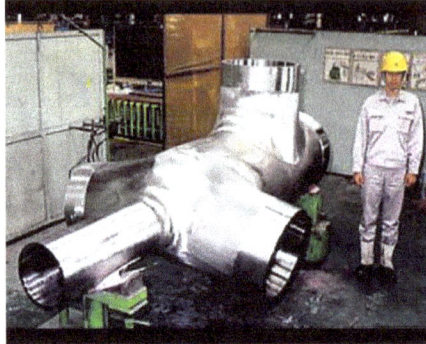

*Fig. 7. PM/HIPP-NNS Offshore Deep-Sea Manifolds.*

**Summary and Conclusion**

The Tera-HIP concept is a viable way forward to manage very large components for the Energy sector, Aerospace or Defense applications. Three different HIP system sizes have been presented. All available for manufacturing and delivery within 24 months after Contract signing. The concept design is based on conventional proven technology used by Quintus Technologies for more than 60 years of operation. The concept is approved by international organizations, i.e. ASME, and the HIP unit can be certified to NADCAP, etc.

# Tailor-Made Net-Shape Composite Components by Combining Additive Manufacturing and Hot Isostatic Pressing

RIEHM Sebastian*[1,a], KALETSCH Anke[1,b], BROECKMANN Christoph[1,c], FRIEDERICI Vera[2,d], WIELAND Sandra[2,e], PETZOLDT Frank[2,f]

[1]RWTH Aachen University, Institute for Materials Applications in Mechanical Engineering, Augustinerbach 4, 52062 Aachen, Germany

[2]Fraunhofer Institute for Manufacturing Technology and Advanced Materials, Wiener Straße 12, 28359 Bremen, Germany

[a]s.riehm@iwm.rwth-aachen.de, [b]a.kaletsch@iwm.rwth-aachen.de, [c]c.broeckmann@iwm.rwth-aachen.de, [d]vera.friederici@ifam.fraunhofer.de, [e]sandra.wieland@ifam.fraunhofer.de, [f]frank.petzoldt@ifam.fraunhofer.de

**Keywords:** HIP, Hot Isostatic Pressing, Additive Manufacturing, LBM, L-PBF, SLM, Composites, Numerical Simulation, Net-Shape

**Abstract.** A promising production route for high quality tailor-made parts can be established by combining Additive Manufacturing (AM) and Hot Isostatic Pressing (HIP): By using a numerical simulation routine, the shape change during HIP can be controlled. These shape-controlled parts are built by Laser Powder Bed Fusion (L-PBF) and consolidated by HIP. After HIP, they exhibit a net-shape geometry that requires only little or even no post-processing at all. In this study, open thin-walled capsules are manufactured by L-PBF, filled conventionally with metal powder, evacuated and sealed and hot-isostatically pressed. Using this processing route, it is possible to combine different materials for the capsule and the powder filling. If capsule and bulk material are identical, the expensive removal of the capsule after HIP can be omitted. By using two different powders, it is possible to produce composite components with a core of high strength and toughness and a wear- or corrosion-resistant surface layer, offering an alternative and competitive production route to conventional HIP cladding. Here three materials are investigated in different combinations: austenitic stainless steel AISI 316L (DIN X2CrNiMo17-13-3), martensitic tool steel AISI L6 (DIN 55NiCrMoV7) and the wear resistant high carbon steel AISI A11 (DIN X245VCrMo8-5-1). A number of technical challenges need to be addressed: the production of dense, thin-walled capsules by L-PBF; L-PBF of carbide rich steels; and controlling the diffusion between corrosion resistant steel and carbon steel. The success of the new process route is demonstrated by metallographic and geometrical investigations.

## Introduction

Hot-Isostatic Pressing (HIP) and Additive Manufacturing (AM) are two process routes to build powder-metallurgical (PM) components. Whereas HIP is a well-established process that has been actively developed for decades, the whole bouquet of beam-based AM processes has only built up momentum over the last years. [1,2]

Combining AM and HIP properly, the benefits of the processes can be exploited and deficiencies can be compensated [3,4]. While AM can be used to produce complex geometries directly from the CAD/CAM process chain in a material-saving manner, production is comparatively time-consuming and thus costly due to the layer structure. The mechanical material properties, which depend significantly on internal defects, lag behind those of conventionally manufactured components. The latter applies in particular to toughness and fatigue strength.

On the other hand, components manufactured by means of HIP have a very fine, isotropic microstructure free of segregation and pores. This makes the mechanical properties comparable to or even better than those of conventionally manufactured components [5]. Due to the conventional capsule design, classical powder HIP is a comparatively complex process, which is limited to simple geometries and small series.

In this study, thin-walled capsules are built by Laser Powder Based Fusion (L-PBF). The actual building time can thus be minimized. Furthermore, L-PBF allows producing complex capsule geometries, which cannot be achieved by conventional welding of sheet metal capsules. In addition, the shape of the capsules prior to construction is optimized with a numerical simulation so that a near-net-shape component is created after HIP.

## Materials and Methods

The aim of this study is to produce complex-shaped fully dense composite components with a corrosion- or wear-resistant outer layer and a tough core built in comparatively short time. Thus, the materials have to be suitable both for L-PBF and for HIP. Table 1 shows the material selection for this study.

*Table 1: Material selection for monolithic and composite components.*

|  | Capsule | Bulk |
|---|---|---|
| Monolithic 1 | Austenitic stainless steel AISI 316L (DIN X2CrNiMo17-13-2) | |
| Monolithic 2 | Martensitic tool steel AISI L6 (DIN 56NiCrMoV7) | |
| Composite 1 | Austenitic stainless steel AISI 316L (DIN X2CrNiMo17-13-2) | Martensitic tool steel AISI L6 (DIN 56NiCrMoV7) |
| Composite 2 | Carbide rich, wear resistant steel AISI A11 (DIN X245VCrMo8-5-1) | Martensitic tool steel AISI L6 (DIN 56NiCrMoV7) |

The "Monolithics" were meant for preliminary tests where capsule and bulk are made from the same material. This article focusses on the "Composites": The capsule is made either from corrosion-resistant 316L or from wear-resistant A11. The tough core consists of L6 steel. Chemical compositions of the steel powders can be found in Table 2. Powder of ferritic steel 430L is used to facilitate crack-free processing of A11 by mixing A11 and 430L.

*Table 2: Chemical composition of used steel powders measured at Fraunhofer IFAM [wt-%].*

|  | C | Si | Mn | P | S | Cr | Ni | Mo | Cu | V | N | B | O | Fe |
|---|---|---|---|---|---|---|---|---|---|---|---|---|---|---|
| L6 | 0.52 | 0.28 | 0.96 | <0.01 | 0.002 | 1.22 | 1.99 | 0.5 | 0 | 0.11 | 0 | 0 | 0 | Bal |
| 316L | 0.021 | 0.8 | 0.95 | 0.009 | 0.004 | 17.3 | 13.2 | 2.5 | 0.05 | 0 | 0.14 | 0.001 | 0 | Bal |
| A11 | 2.45 | 0.98 | 0.42 | 0.02 | 0.087 | 5 | 0 | 1.49 | 0 | 8.28 | 0.098 | 0 | 0.035 | Bal |
| 430L | <0.018 | 0 | 0 | 0 | 0 | 16–18 | 0 | 0 | 0 | 0 | 0 | 0 | 0 | Bal |

The experiments regarding L-PBF were conducted at Fraunhofer IFAM (Bremen/Germany) with an L-PBF unit "M270 Dual Mode" from EOS GmbH. This unit is equipped with an extended powder bed heating option of up to 300 °C.

Hot Isostatic Pressing – HIP'17                                    Materials Research Forum LLC
Materials Research Proceedings **10** (2019) 203-209        doi: http://dx.doi.org/10.21741/9781644900031-27

In order to find suitable parameters for the L-PBF process, a multi-step parameter variation study was done. At first, by variation of scan velocity and laser power and thus by variation of energy density, thin lines and walls and small cubes were produced. In a second step, an appropriate parameter set was fine-tuned by building overhangs and different angles. Finally, the HIP capsules used in this study could be built successfully. [6]

The HIP cycles were performed at IWM of RWTH Aachen University using a HIP unit "Shirp 20/30-200-1500" by ABRA Fluid AG (Widnau/Switzerland). Prior to HIP, the capsules were prepared for the cycle. They were cleaned with isopropyl alcohol and a long filling pipe was TIG-welded to the L-PBF-built socket. A minimum quantity of powder was calculated as the product of capsule volume and tap density. At least this minimum quantity was filled into the capsules using a vibration table. After filling, the capsules were evacuated and sealed.

As the results of the study should be transferable into industrial applications as seamless as possible, HIP parameters were chosen which are considered standardized industrial parameters: $T_{HIP}$ = 1125 °C, $p_{HIP}$ = 110 MPa and $t_{Hold}$ = 3 h. The pressure was applied starting at 750 °C.

**Results & Discussion**

*L-PBF of Capsules.* The capsule design used for the investigations is shown in Figure 1: The technical drawing on the left illustrates the dimensions. The wall thickness, which is specified as 2 mm, is variable: Capsules with 1.0 mm, 1.5 mm and 2.0 mm wall thickness have been built. The experiments showed that using L-PBF, a minimum wall thickness of 1.0 mm is required in order to build gastight walls.

The wall thickness of 1.5 mm in the upper part of the capsule according to Figure 1 was not varied. This thickness was necessary to join a filling and evacuation tube by means of TIG welding.

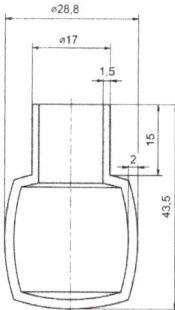

*Figure 1: Capsule design. Left: capsule geometry, right: L-PBF-made capsules of 316L.*

Capsules made of 316L and L6 could be successfully built with several combinations of L-PBF process parameters [6].

For A11 however, the L-PBF-building turned out to be challenging. The material forms martensite when it cools down because of the high C content. As martensite is brittle and leads to high stresses, cracks and delaminations occur after L-PBF. Two possible solutions were examined:

Martensite start temperature of A11 is about 230 °C to 300 °C. Using a heating system which is capable of heating the building plate to 300 °C in order to build above martensite start

Hot Isostatic Pressing – HIP'17                                    Materials Research Forum LLC
Materials Research Proceedings **10** (2019) 203-209         doi: http://dx.doi.org/10.21741/9781644900031-27

temperature was not successful either: Cubes with a height of about 5 mm could be built. However, they showed cracks, some of which reached almost through the entire component.

The second approach was to develop an alloy that makes it possible to process the high-carbide steel. A11 was mixed with different ratios of 430L. Using a 430L content of over 17.5 wt-% and pre-heating to 300 °C, cubes with only few cracks could be produced (Figure 2, left). The same result could be reached for a mixture of 50 wt-% 430L and using 80 °C substrate temperature (Figure 2, middle).

A major disadvantage of the powder mixture is a loss in hardness, see Figure 2, right. Since the capsules are intended to serve as wear protection, the hardness must be sufficient for the application. Whether the hardness loss can be accepted must be decided on a case-by-case basis. Developing a HIP can of wear resistant material, such as A11 through AM, remains a very promising task that still needs a solution.

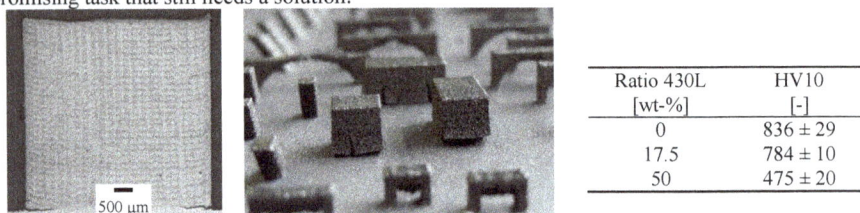

| Ratio 430L [wt-%] | HV10 [-] |
|---|---|
| 0 | 836 ± 29 |
| 17.5 | 784 ± 10 |
| 50 | 475 ± 20 |

*Figure 2: Nearly crack-free cube of A11 + 17.5 wt-% 430L (left); small parts of A11 + 50 wt-% 430L (middle); hardness HV10 of A11 and mixtures (right).*

*HIP of Capsules and Numerical Finite Element Simulation.* All L-PBF-built capsules could be successfully compacted by HIP to full density. When the wall thickness was below 1 mm, the capsules could not be densified successfully. Obviously the capsules were not gastight.

Using a numerical simulation routine that was developed by IWM it is possible to predict shrinkage and densification during HIP [7]. For this simulation, the capsules were modeled in the FEM software Simulia Abaqus and provided with input parameters: certain material parameters and relative powder density from filling. A homogenous density distribution was assumed. Figure 3 shows a comparison between the real HIPed and the simulated capsule shape of a capsule where capsule and bulk are made of 316L.

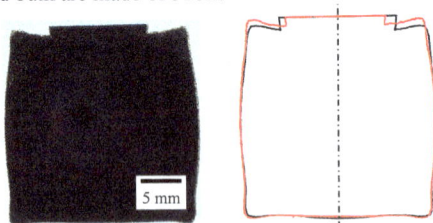

*Figure 3: Outline of real HIPed vs. simulated capsule shape: capsule/bulk 316L.*

*Metallographical Investigation.* In Figure 4 two benefits of the HIP treatment are illustrated for A11 and for 316L respectively: For each material, on the left side a component as-built, i.e. after L-PBF, is shown. A number of pores and voids are visible and structures similar to welding beads, that are artifacts of the L-PBF building. On the right side, the condition after HIP is shown: All inner pores are closed. Furthermore, the L-PBF structure is vanished. Obviously, HIP resulted in a macroscopic homogenization of the material. For 316L after HIP, the grain boundaries were highlighted by etching.

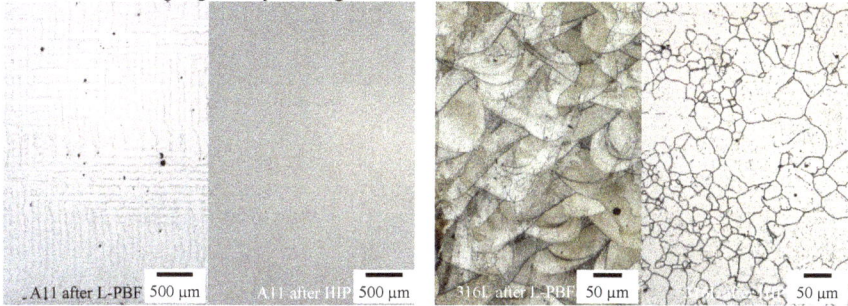

*Figure 4: Optical light microscopy images before and after HIP of A11 (left) and 316L (right).*

The structure after HIP (capsule and filling: 316L; wall thickness: 1 mm) is shown in Figure 5. The metallographic cross section was prepared by grinding, polishing and etching according to Beraha III. The different colouring from light blue to orange-brown may be misleading: It does not allow any conclusions to be drawn about the microstructure.

Although different 316L powder grain sizes were used for capsule (grain size: -53 +20 µm; D50 = 46 µm) and filling (grain size: -500 +0 µm; D50 = 197 µm), the grain sizes no longer differ after HIP. HIP obviously causes recrystallization in capsule and bulk, thus generating evenly sized grains.

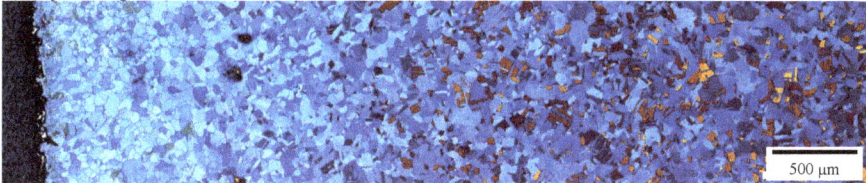

*Figure 5: Microstructure after HIP: capsule 316L, bulk 316L.*

Figure 6 shows a cross section of a composite capsule after HIP, where the walls are made of 316L and the filling of L6. At the outside of the wall, some open porosity is clearly visible, reaching up to 200 µm inwards. This jagged surface must be appropriately considered in industrial applications. The interface between these two materials can be well seen, though there is no separation or voids. The quality of the HIP bond still needs a special evaluation

*Figure 6: Microstructure after HIP: capsule 316L, bulk L6.*

A higher magnification of the interface between capsule and bulk can be seen in Figure 7. The numbers indicate the approximate locations where EDX measurements were made. The diagram shows how the content of chromium and nickel decreases at the transition from the capsule to the inside. Apparently, a diffusion zone of 20 μm to 30 μm is formed.

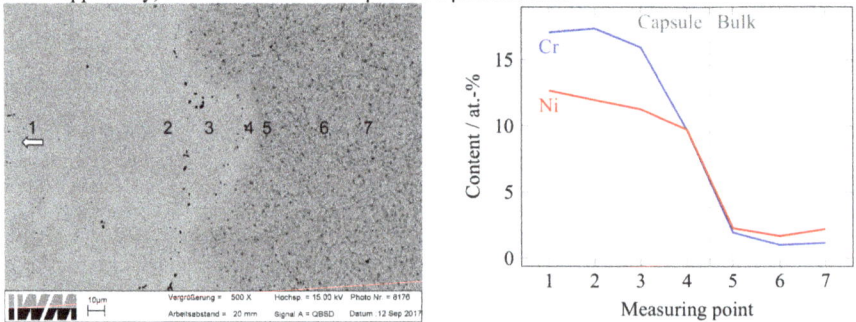

*Figure 7: EDX at intersection between capsule (316L) and bulk (L6).*

**Summary and Outlook**

The combination of AM and HIP can be used to easily build complex and irregular-shaped capsules. A hollow capsule is made of wear- or corrosion-resistant steel and filled with a tough steel powder. 316L and L6 can be well manufactured via L-PBF. In order to process the high-carbide steel A11, further parameter investigations must be carried out.

First microscopic examinations of the capsules produced in this way after HIP seem promising. Pore-free compaction was achieved and the materials bonded well. In further investigations, the components must be qualified in comparison to conventionally manufactured components, especially regarding key features such as hardness of A11 and bonding of capsule and bulk.

**References**

[1] H.V. Atkinson, S. Davies, Fundamental aspects of hot isostatic pressing: An overview, Metall and Mat Trans A 31 (2000) 2981–3000. https://doi.org/10.1007/s11661-000-0078-2

[2] C.Y. Yap, C.K. Chua, Z.L. Dong, Z.H. Liu, D.Q. Zhang, L.E. Loh, S.L. Sing, Review of selective laser melting: Materials and applications, Applied Physics Reviews 2 (2015) 41101. https://doi.org/10.1063/1.4935926

[3]  A. Kumar, Y. Bai, A. Eklund, C.B. Williams, Effects of Hot Isostatic Pressing on Copper Parts Fabricated via Binder Jetting, Procedia Manufacturing 10 (2017) 935–944. https://doi.org/10.1016/j.promfg.2017.07.084

[4]  H. Hassanin, K. Essa, C. Qiu, A.M. Abdelhafeez, N.J.E. Adkins, M.M. Attallah, Net-shape manufacturing using hybrid selective laser melting/hot isostatic pressing, Rapid Prototyping Journal 23 (2017) 191. https://doi.org/10.1108/RPJ-02-2016-0019

[5]  H. Masuo, Y. Tanaka, S. Morokoshi, H. Yagura, T. Uchida, Y. Yamamoto, Y. Murakami, Effects of Defects, Surface Roughness and HIP on Fatigue Strength of Ti-6Al-4V manufactured by Additive Manufacturing, Procedia Structural Integrity 7 (2017) 19–26. https://doi.org/10.1016/j.prostr.2017.11.055

[6]  V. Friederici, S. Wieland, Laser Beam Melting Of Heat-treatable 56NiCrMoV7 And Wear-resistant FeCrV10 Steel, in: European Powder Metallurgy Association (Ed.), Euro PM2017 Proceedings, 2017.

[7]  C. Van Nguyen, Numerical Simulation of Hot Isostatic Pressing with Particular Consideration of Powder Density Distribution and Temperature Gradient, 1st ed., Shaker, Herzogenrath, 2016.

# Efficient Modeling of PM HIP for Very Large NNS Parts (up to 2.5 Meter Diameter) and Key Physical, Material and Technological Parameters to Control Dimensional Scattering in a 15 mm Range

Dr. Gerard Raisson[1,a,*], Prof. Vassily Goloveshkin[2,b], Dr. Anton Ponomarev[2,c], Andrey Bochkov[3,d], Yuri Kozyrev[3,e]

[1]Consultant, 63800 Cournon d'Auvergne, France

[2]Moscow Technological University, IPRIM RAS, Moscow Russia

[3]Moscow Technological University, Moscow, Russia

[a]gerard.raisson@gmail.com, [b]nikshevolog@yandex.ru, [c]avpon@yandex.ru, [d]andrey.bochkov@gmail.com, [e]gmile88@mail.ru

**Keywords:** HIP Modeling Data Identification, HIP Dilatometer

**Abstract:** To get via PM HIP a 2.5 meter diameter or even larger part in a Near Net Shape (NNS) configuration at the first attempt with a 15 mm max over thickness is a very challenging task. However, its solution will enable production of such parts for various critical applications. To achieve this goal, it is necessary to increase precision of HIP modeling. Our analysis has shown that rather than to work on constitutive equations, it is more efficient to improve the consistency of the data base of rheological properties for the powder and capsule materials and more particularly at the first step of HIP cycle which controls heat conductivity. Also it is necessary to account additional effects of the strain rate hardening for the capsule material and of the initial packing density for the powder.

Independently of modeling, it is necessary to control all parameters generating the dimensional scattering (HIP cycle trajectory, temperature homogeneity, filling and handling of capsules). Modeling helps to define what the most critical parameters are and what dimensional tolerances can be respected.

## Introduction

Production of very large parts by HIPing of thin wall capsules filled with pre-alloyed powder is one of the major fields of development of powder metallurgy where this technology can be advantageous. For large parts, economically speaking over thickness is an important issue because of the cost of scrapped materials and subsequent machining costs. Limitation of over thickness to 15 mm means a control of distortion of +- 7.5mm. For a 2.5 m diameter part if we consider isotropic conditions, the shrinkage will be around 15%, i.e. 375 mm. So, 7.5 mm precision requirement represents 2% of total distortion.

To reach this very ambitious target, modeling is an important tool. The paper shows what are the key points and difficulties with modeling. Besides modeling, a lot of parameters (process or mechanical) can be a cause of scattering. For example, control of filling density and its uniformity is not at all trivial for thin wall large parts. Modeling can also be very useful to evaluate sensitivity of final geometry to these technological parameters.

## Modeling

*Mechanical analysis*
Development of efficient modeling programs has been carried out in the beginning of 1980s [1] and [2]. Generally speaking, they use Green analysis [3] using stresses equilibrium with:
- Full dense material flow stress
- 2 coefficients depending of relative density giving effect of density on densification related to pressure and on shear distortion related to shear stress.

Some other authors [4], [5] have proposed to introduce other coefficients to take in count crossed effect of pressure and shear stress.
Later on, we will use the Green formulation:

$$\frac{\sigma^2}{f_2^2} + \frac{T^2}{f_1^2} = Y$$

Where T is the shear stress, σ the mean stress (or pressure), Y full dense material flow stress and $f_1$ and $f_2$ two coefficients (plasticity functions) depending on the relative density of powder during densification.

*Parameters identification*
In order to identify these functions and coefficients mechanical trials are performed:
- Full dense material flow stress (Y in equation 1) is measured by classical upsetting at several temperatures with a controlled strain rate.
- Pressure coefficient ( $f_2$ function) can be obtained either through interrupted HIP cycles. In any case a calculation is needed to know the actual pressure in the powder that is shielded by a HIP can.
- Shear stress coefficient ( $f_1$ function) can be obtained by upsetting trials with several density levels.
- In the frame of this presentation, it has been used HIP dilatometer technology [4] and [5]. The principle of device is given in figure 1. An axial and a radial recording have been carried out with the same HIP cycle. The main advantage of this technology is to allow a continuous recording of data. Shielding effect of capsule has to be calculated to get actual pressure and shear stress in powder. In adjusting thickness and material of capsule, it is possible to control shear stress and this parameter is closer of conditions met in an actual industrial capsule in comparison of conventional upsetting. Shielding effect of capsule is relatively easy to calculate by an analytical way for a long thin walled capsule.
- In this paper it will be presented mainly $f_2$ function of density results. Presentation of $f_1$ function will be the subject of a future publication.

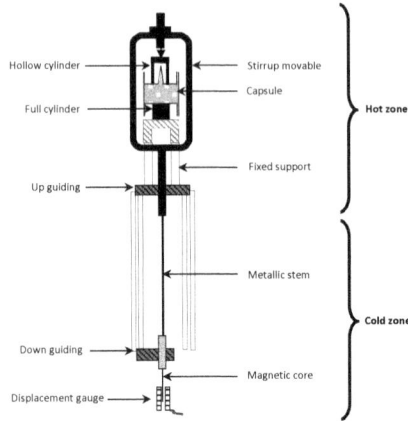

*Figure 1: Principle of Villetaneuse HIP dilatometer [6] and [7].*

A first comment, at this level, is that the principle of full dense material rheology is not actually known and, some way:

- Microstructure is evolving all along HIP cycle with associated effect on flow stress.
- Strain rate at the scale of particle is inhomogeneous particularly in the first stage of densification.

Actually, it means that it is not realistic to identify independently full dense material flow stress and coefficients. In fact, only a set $(Y, f_1, f_2)$ is identified and valid physically speaking. For example, for isotropic trial, if full dense material flow stress is given, identification gives a set $(Y, f_2)$ in order that $P = f_2 * Y$ where P is the (calculated) actual pressure in the powder. If there is a Y misestimating, $f_2$ value will compensate it if we give the sets of values. It means that it is necessary for identification trials (interrupted cycles or HIP dilatometer) to stay close to the HIP cycle that is used for making a part and to carry out upsetting trials at the same temperature of cycle interruptions. We'll see in the next paragraph what means:"to stay close to the HIP cycle".

Three capsules of Ti6Al4V PREP powder have been densified according to 3 different pressure ramps when heating rate is kept constant (figure2, [6], [7]). Pressure in powder is also indicated. Figure 3 gives corresponding densification curves.

Using the same Y function of temperature law for the full dense material, $f_2$ function of relative density is obtained and presented in figure 4. It can be seen that in a certain validity domain, a simple rheology of "full dense material" can be used. The same result has been obtained on 316L and nickel base grades.

On figure 5, it is shown densification rate in function of relative density for 316L with several pressure levels with the heating rate kept constant[7]. Densification is over before the temperature dwell is reached and the curves are identical taking in account effect of initial density. Same trials were carried out for TI6AL4V and nickel base grade with the same results. It means that the deformation pattern during densification is similar, what could justify proposed approach. It is also evident that a change in heating rate will change these curves and the sets ($Y$, $f_2$) should be adjusted.

The important question is: what precision of parameters involved in mechanical analysis (rather sets of parameters) is needed? For massive parts, shear stress level in powder is low as the influence of capsule is small. Therefore, some incertitude on $f_1$ coefficient could be acceptable.

In opposite, particularly due to the heterogeneity of temperature (as it will be seen later), the densification curve (the $f_2$ coefficient and the Y function of temperature) has to be as precise as possible.

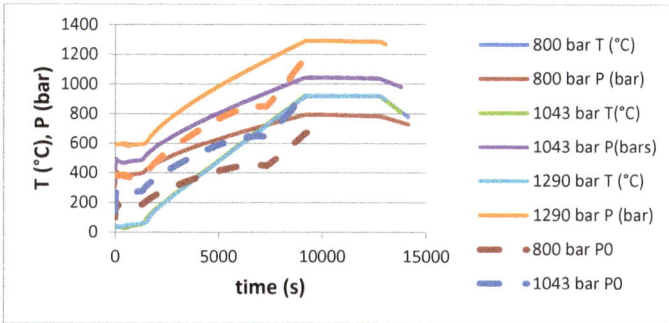

Figure 2: TI6AL4V HIP parameters with calculated powder pressure.

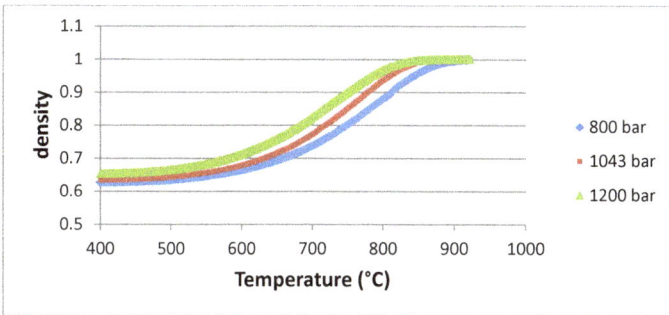

Figure 3: TI6AL4V densification curves.

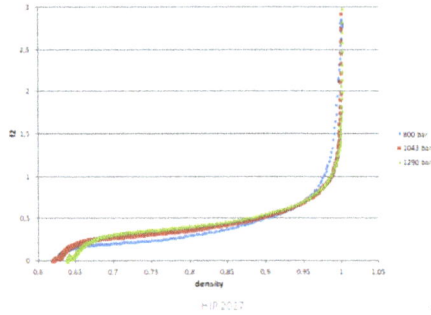

*Figure 4: $f_2$ function of density.*

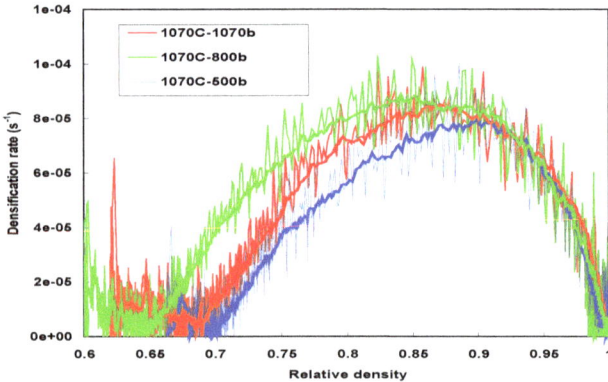

*Figure 5: 316L densification rate function of relative density for three pressure levels [7].*

Also the strain rate sensitivity of the capsule material can influence its stiffness and the resulting deformation pattern in the HIP cycle. Therefore we have undertaken the following Strain Rate study.

To study the influence of the strain hardening of the capsule material on the HIP deformation pattern and the final geometry we consider the deformation of a long cylinder (the length much exceeding the radius).

For such a HIP can we can neglect the influence of the lids.

With Z being the axis of symmetry, let the area $0 < r < R, 0 < z < H$ be the area of powder and $R < r < R + h, 0 < z < H$ - of the capsule material

As before, the Green's criterion was used to describe the "Plasticity Condition" for the powder material:

$$\frac{\sigma^2}{f_2^2} + \frac{s^2}{f_1^2} = Y^2 \tag{1}$$

Here: $\sigma$ - the average stress; $s^2$- intensity of the deviatoric stress tensor; $f_1, f_2$ - experimental values of the plasticity functions of the current density $\rho: Y$ - the yield strength of the full dense powder material.
Here:

$$\varepsilon_{ij} = \omega \frac{\partial \Phi}{\partial \sigma_{ij}} \tag{2}$$

where: $\varepsilon_{ij}$ - the components of the strain rate tensor,
The capsule material is non-compressible and is described by the ideal plasticity law:

$$s^2 = T^2 \tag{3}$$

Where $T$ - is the yield strength.
Let $U_r$, $U_z$ be the radial and axial rates. We consider the strain rate on the symmetry axis constant. Then in the powder material:

$$U_z = -\varepsilon z, \; U_r = -Ar. \tag{4}$$

Due to the non-compressibility of the capsule material and continual radial rate at $r = R$ we have:

$$U_z = -\varepsilon z, \; U_r = \frac{1}{2}\varepsilon r - \frac{1}{2}\varepsilon \frac{R^2}{r} - A\frac{R^2}{r} \tag{5}$$

Using the equilibrium equation of the cylindrical system of coordinates:

$$\frac{\partial \sigma_r}{\partial r} + \frac{\sigma_r - \sigma_\varphi}{r} = 0, \tag{6}$$

and an integral equilibrium equation relative to the $Z$ axis, we get:

$$2\pi \int_0^{R_1} \sigma_z r dr = -P\pi R_1^2 \tag{7}$$

(where $R_1 = R + h$, $P$ - external pressure), assuming that $h << R$ (relatively thin capsule), we obtain the following relations for $\varepsilon$, $A$

$$Y \cdot \frac{f_1^2 (3\varepsilon - 1)}{\sqrt{\left(9 f_2^2 - 2 f_1^2\right) + 3 f_1^2 \left((1 - \varepsilon)^2 + 2\varepsilon^2\right)}} + 2 \frac{h}{R} T \frac{\sqrt{3\varepsilon}}{\sqrt{3\varepsilon^2 + 1}} = 0 \tag{8}$$

$$A = \frac{1}{2}(1 - \varepsilon) \tag{9}$$

Notice that $\varepsilon \in \left[0; \dfrac{1}{3}\right]$:

at $f_1 \to 0$ the value of $\varepsilon \to 0$ - flat deformation

at $f_2 \to \infty$ the value of $\varepsilon \to 0$ - flat deformation.

at $h \to 0$ the value of $\varepsilon \to \dfrac{1}{3}$ - uniform compression

This partially explains the difficulties of predicting the initial stages of densification. One the one hand, the capsule is thin, on another -at the low densities the value of $f_1$ is small. As far as $\varepsilon \in \left[0; \dfrac{1}{3}\right]$, i.e. relatively small, in accordance with (8), we have an approximate equation to determine the value of $\varepsilon$

$$\varepsilon = \frac{Y}{\left\{ Y \cdot \dfrac{3 f_1^2}{\sqrt{\left(9 f_2^2 - 2 f_1^2\right)}} + 2\sqrt{3} \dfrac{h}{R} T \right\}} \cdot \frac{f_1^2}{\sqrt{\left(9 f_2^2 - 2 f_1^2\right)}} \tag{10}$$

The values of $H$ and $R$ as functions of the density $\rho$ are determined by (11) and (12) :

$$H(\rho) = H_0 \exp\left\{ -\int_{\rho_0}^{\rho} \frac{Y}{Y \cdot \dfrac{3 f_1^2}{\sqrt{\left(9 f_2^2 - 2 f_1^2\right)}} + 2\sqrt{3} \dfrac{h}{R} T} \cdot \frac{f_1^2}{\sqrt{\left(9 f_2^2 - 2 f_1^2\right)}} \frac{d\rho}{\rho} \right\} \tag{11}$$

$$R^2(\rho) = R_0^2 \frac{\rho_0}{\rho} \cdot \frac{H_0}{H} \tag{12}$$

Where $R_0$, $H_0$ are the initial values.

If the capsule yield strength as a function of the strain rate is presented as: $T = B(e_u)^\alpha$ (13)

Where $e_u = \dfrac{3}{2}\sqrt{\left(\varepsilon_r^2 + \varepsilon_\varphi^2 + \varepsilon_z^2\right)}$

Let the external pressure as a function of times is:

$$P = P_0 t \qquad (14)$$

Then, introducing the following dimensional values:

Dimensional time $\tau$ as $t = \dfrac{Y}{P_0}\tau$; dimensional $A$ as $\overline{A}$ in the form of

$A = \dfrac{P_0}{Y}\overline{A}$ (the over score further omitted), get the following relations:

$$\frac{2f_1^2\left[1-x\right]}{\sqrt{\left(9f_2^2 - 2f_1^2\right)\left(2+x\right)^2 + 6f_1^2\left(2+x^2\right)}} - \gamma A^\alpha\left(x^2+x+1\right)^{\frac{\alpha}{2}}\frac{\sqrt{3}x}{\sqrt{\left(1+x+x^2\right)}}\frac{h}{R} = 0 \quad (15)$$

$$\frac{1}{3}\frac{\left[9f_2^2(2+x)+2f_1^2(1-x)\right]}{\sqrt{9f_2^2\left(2+x\right)^2 + 2f_1^2\left(1-x\right)^2}} + \gamma A^\alpha\left(x^2+x+1\right)^{\frac{\alpha}{2}}\sqrt{\frac{1}{3}}\frac{\left(x+2\right)}{\sqrt{\left(1+x+x^2\right)}}\frac{h}{R} = \tau \quad (16)$$

Where $\gamma = \dfrac{B}{Y}\left(\dfrac{3\sqrt{2}}{2}\dfrac{P_0}{Y}\right)^\alpha$, $x = \dfrac{\varepsilon}{A}$

The density is described by the equation:

$$\frac{d\rho}{d\tau} = \rho A(2+x) \qquad (17)$$

The solution for a specific material is defined by the two parameters:

$\gamma = \dfrac{B}{Y}\left(\dfrac{3\sqrt{2}}{2}\dfrac{P_0}{Y}\right)^\alpha$, characterizing the rate of loading and

$\alpha$ - that describes the yield strength dependence of the strain rate

Calculations were done for the $f_1$ and $f_2$ coefficients derived from the dilatometric studies and show a serious influence of the strain rate dependence of the capsule flow stress on the deformation pattern measured by anisotropy of deformations (figure 6).

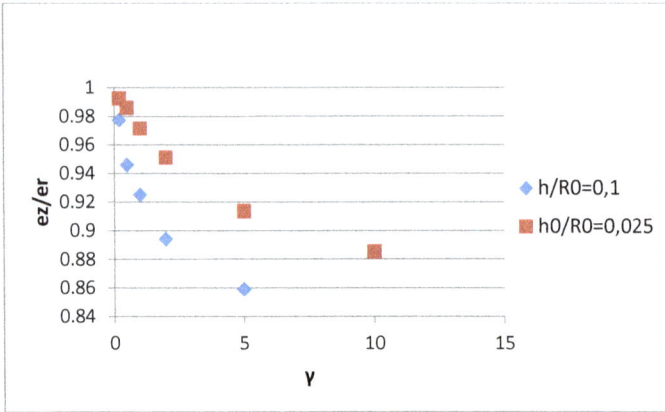

*Figure 6: Anisotropy of densification in function of γ parameters (α 0.2).*

This parametric study shows that actual strain rate dependence of capsule flow stress has to be accounted as it can strongly influence the deformation pattern:
- Densification rate varies along HIP cycle
- α and γ vary along HIP cycle
- anisotropic effect grows when temperature is growing, strain-rate is lowering and capsule thickness-diameter ratio is growing.

*Thermal conductivity:*
For massive parts, it is mandatory to introduce thermal conductivity because of non negligible temperature gradients in the part.

Thermal conductivity has been introduced in LNT and Abouaf modeling [1], [9]. An important point is to know thermal conductivity as a function of relative density. Using work of

$$\lambda(\rho) = \lambda_0 * \left( \frac{\rho - \rho_0}{1 - \rho_0} \right)^{1,46*(1-\rho_0)}$$

Argento [8] , the relationship is:                                             where $\lambda_0$ is thermal conductivity of full dense material, $\rho_0$ initial density of powder, $\lambda(\rho)$ thermal conductivity at density $\rho$. Figure 7 shows the curves $\lambda$ f($\rho$) for several initial densities function. The difference in the values of the thermal conductivity in the same points of the HIP trajectory is significant and must be properly accounted. During the very beginning of densification is due to initial cold pressurization of HIP unit, so, strain of powder is purely elasto-plastic. Relationship between pressure and density can be obtained through Ashby or Raisson analysis [10], [11]. These very close results are given in figure 8.

Hot Isostatic Pressing – HIP'17                          Materials Research Forum LLC
Materials Research Proceedings **10** (2019) 210-223        doi: http://dx.doi.org/10.21741/9781644900031-28

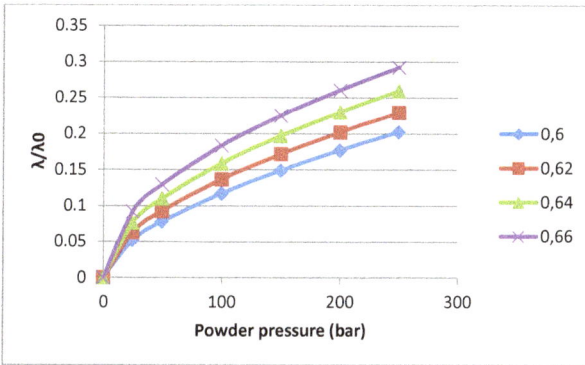

*Figure 7: Thermal conductivity of TI6AL4V in function of pressure and initial density.*

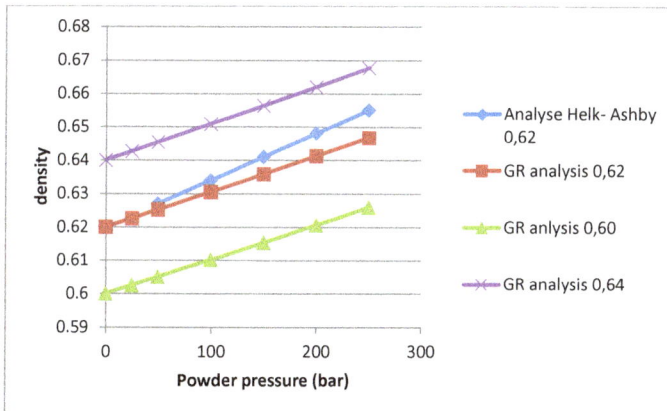

*Figure 8: Density of TI6AL4V powder in function of pressure.*

It is evident that there is a strong effect of initial density and pressure on thermal conductivity during the first stage of HIP cycle. At higher density, the effect becomes negligible.

*Critical parameters for dimensional scattering:*
As far as the massive parts are concerned, two parameters are critical for precision of modeling:

**a.**   Filling density and its possible heterogeneity in a large (high) can where the weight effects become significant:

i.    Mean filling density knowing that 1% difference of density generates 0.3% geometry discrepancy and mainly

ii.   Heterogeneity of filling density which generates local discrepancy, the worst being 3D effect (bending) when heterogeneity is asymmetric [10]. For large thin walled capsule it is difficult to get the same density in any place. It needs a specific know-how to be developed.

    iii.      Additional complement of tapping during the transportation between filling and HIP shops.

**b.**    Taking into account thermal conductivity and consequently again filling density and HIP cycle and particularly the pressure and temperature ramp and in order to estimate the relative effect of parameters, two modeling have been carried out:

    i.      For a low alloyed steel ring (internal diameter 183 mm, external diameter 1100 mm, height 100 mm) shrinkage ratio of internal diameter has been calculated in function of initial thermal conductivity (figure 9). The effect is major and at low initial thermal conductivity, the deformation is even negative (outward)

    ii.      For a nickel base massive ring (internal radius 500 mm, external diameter 1000 mm, height 500 mm) change of external diameter and height has been calculated in function of filling density with two hypothesis:

        1.    Simple law giving initial thermal conductivity function of filling density.

        2.    Calculation of density and thermal conductivity function of filling density and pressure after pressurization according to **b-c** paragraph.

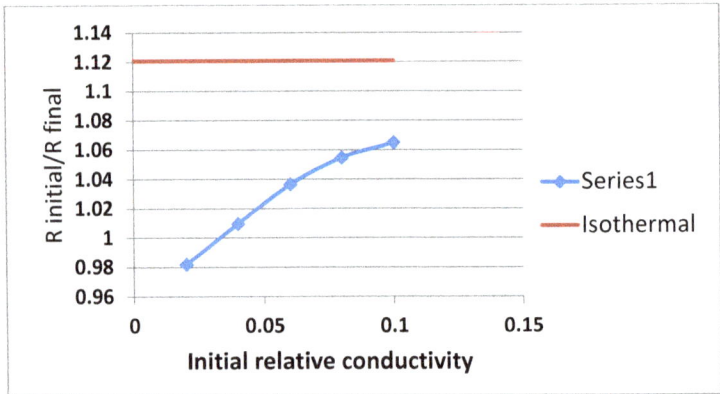

*Figure 9: Internal diameter shrinkage ratio of low alloyed steel ring function of initial thermal conductivity.*

Hot Isostatic Pressing – HIP'17                     Materials Research Forum LLC
Materials Research Proceedings **10** (2019) 210-223     doi: http://dx.doi.org/10.21741/9781644900031-28

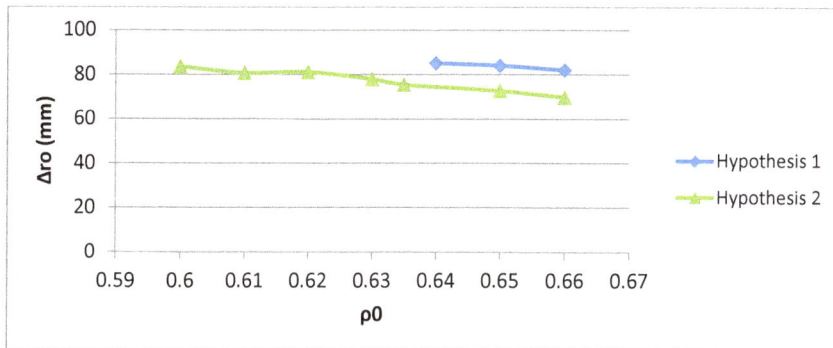

Figure 10: Variation of external radius of a nickel base ring function of filling density and
hypothesis for treatment of density and thermal conductivity during pressurization step.

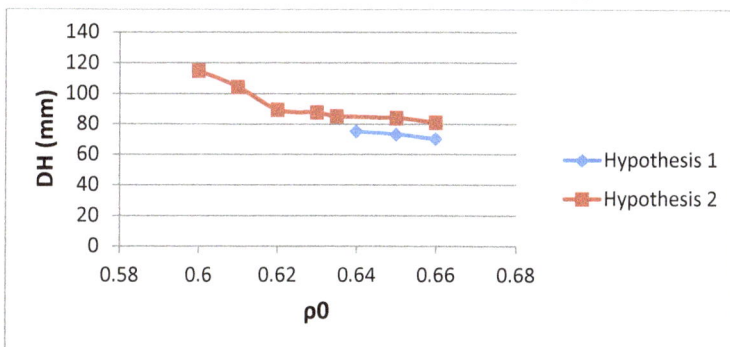

Figure 11:   Variation of height of a nickel base ring function of filling density and hypothesis for
treatment of density and thermal conductivity during pressurization step.

The effect (figures 10 and 11) is minor but not negligible (around 10 mm) and should be
accounted for large parts

*Others scattering parameters:*

    a.  Geometry of the capsule:
        i.  At the level of making of large size welded capsule
        ii.  During filling due to stresses generated by the weight of powder. Some
            additional supporting tooling needs to be developed.
        iii.  During handling of a several tons part. If handling points are not well
            designed, there is a risk of irreversible plastification of capsule.
        iv.  During handling again due to shocks giving depressions. Experience
            shows that it is not a negligible risk.

b.  Cooling step: particularly for low alloyed steel (when massive parts of the HIP capsule are made from this steel), γ->α transformation corresponds to up to 0,7 % expansion as a function of cooling rate. Due to non-uniform temperature distribution, it generates some anisotropic plastic deformation of the HIP capsule reaching several millimeters.

c.  Powder batch: the knowledge of powder characteristics is an evident issue.  It means that filling density and rheology are perfectly reproducible. On figure 12, it is shown densification curve for Astroloy powder produced by Rotating Electrode Pulverization and Gas Atomization with the same HIP cycle and particle size distribution [11]. Effect of process which controls microstructure is strong. But the most important item is actual filling density.

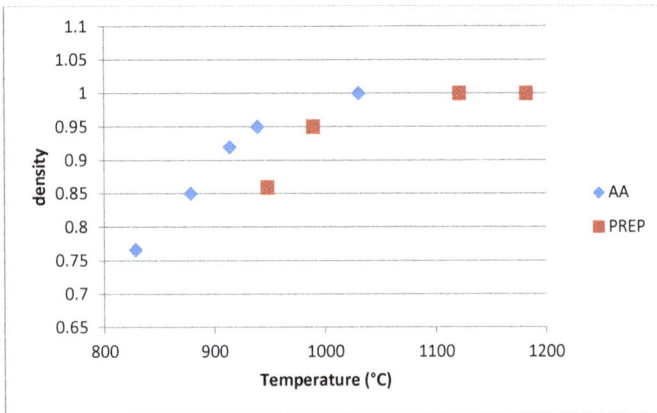

*Figure 12: Densification curves for Astroloy PREP and AA powders. [13].*

**Conclusion**

To get the given target to have at disposal a modeling allowing at the first attempt to produce a 2.5 m diameter massive part with a precision better than +-7.5 mm seems realistic:

-   It is not necessary to use a sophisticated law for full dense material rheology as far as heating rate is rigorously controlled in production HIP cycles and pressure ramp kept in a validity domain.
-   Initial conditions are fundamental issues to control thermal conductivity. Some progress is still necessary to have a good physical description of the first step of densification (pressurization).
-   Modeling is not enough to guaranty to reach the target. A lot of other parameters have to be taken in account. Modeling can help to determine the order of magnitude of their effect.

**References**

[1] Abouaf, Chenot, Raisson: Finite Elements Simulation of Hot Isostatic Pressing of Metal Powder. J.Numerical Methods in Engineering 25, 191-212, 1988. https://doi.org/10.1002/nme.1620250116

[2] Samarov et al: Application of HIP for Near Net Shape critical parts and components. Hot Isostatic Pressing 1993 L.Delaey and H.Tas (editors) 1994 Elsevier Science B.V. pp 171-185. https://doi.org/10.1016/B978-0-444-89959-0.50022-0

[3] R.J.Green: Int.Mech.Sci., 1972, 14, (4), 215-224. https://doi.org/10.1016/0020-7403(72)90063-X

[4] Sanchez, Ouedraogo, Stutz, Dellis: Integration of a new viscoplastic constitutive equation in a finite element code. Powder Metallurgy World Congress, 98, Vol 2, pp 531-536, Granada, Spain.

[5] P.Viot: Modelisation du comportement viscoplastique des poudres métalliques. Developpement d'un nouveau dispositif experimental de Compression Oedometrique à Chaud. *« Modeling of viscoplastic behavior of metallic powder. Development of a new Hot Oedometric Pressing device »*. PHD thesis 2000 december 18 Grenoble 1.

[6] Fondere, Federzoni, Raisson: Identification Models of Powder Densification using in-situ HIP Dilatometer. 2004 PM World Congress. Vienna October 2004.

[7] Burlet, Gillia : Model identification for powder densification. HIP 2005 Paris may 2005

[8] C.Argento and D.Bouvard: Modeling the effective thermal conductivity of random packing of spheres through densification. *Int. J. Heat Mass Transfer* vol.**39**-7, 1343-1350 (1996)

[9] Goloveshkin, Raisson, Ponomarev, Bochkov: Accounting of non-stationary Temperature Field while modeling of HIP for large size Components. HIP 2011 Kobe April 2011.

[10] Samarov, Raisson, Seliverstov, Goloveshkin: Principles of shape and densification Control during HIP. HIP 2011 Kobe April 2011.

[11] Ashby: The Modelling of Hot Isostatic Pressing. Proceedings of th international Conference on Hot Isostatic Pressing. Lulea, 1987.

[12] Van Nuguyen, Bezold, Broeckmann: Anisotropic Shrinkage of Hot Isostatically Pressed Components. HIP 2014 Stockholm June 2014.

[13] Than-Chi Lu: Structure superficielle des poudres de superalliages base nickel et mécanismes intervenant au cours de la densification. *Nickel base Superalloys Surface Structure and involved mechanism during densification"* PHD thesis Ecole des Mines de Paris 1987.

Hot Isostatic Pressing – HIP'17                                          Materials Research Forum LLC
Materials Research Proceedings 10 (2019) 224-234          doi: http://dx.doi.org/10.21741/9781644900031-29

# Small Modular Reactor Vessel Manufacture/Fabrication Using PM-HIP and Electron Beam Welding Technologies

David W. Gandy[1,a*], Craig Stover[1,b], Keith Bridger[2,c], Steve Lawler[2,d], Matt Cusworth[2,e], Victor Samarov[3,f] and Charlie Barre[3,g]

[1]Electric Power Research Institute, 1300 W.T. Harris Blvd, Charlotte NC, USA

[2]University of Sheffield, Nuclear Advanced Manufacturing Research Centre, Advanced Manufacturing Park, Brunel Way, Rotherham, UK

[3]Synertech-PM, 11711 Monarch St, Garden Grove, CA, USA

[a]davgandy@epri.com, [b]cstover@epri.com, [c]k.bridger@bridgerweldingengineering.com
[d]steven.lawler@namrc.co.uk, [e]matthew.cusworth@namrc.co.uk
[f]victor@synertechpm.com, [g]charlie@synertechpm.com

**Abstract.** Many of the same manufacturing/fabrication technologies that were employed for light water reactors (LWR) plants built 30-50 years ago are also being employed today to build advanced light water reactors (ALWRs). Manufacturing technologies have not changed dramatically for the nuclear industry even though higher quality production processes are available which could be used to significantly reduce overall component manufacturing/fabrication costs. New manufacturing/fabrication technologies that can accelerate production and reduce costs are vital for the next generation of plants (Small Modular Reactors (SMR) and GEN IV plants) to assure they can be competitive in today's and tomorrow's market.

This project has been assembled to demonstrate and test several of these new manufacturing/ fabrication technologies with a goal of producing critical assemblies of a 2/3rds scale demonstration SMR reactor pressure vessel (RPV). Through use of technologies including: powder metallurgy-hot isostatic pressing, (PM-HIP), electron beam welding, diode laser cladding, bulk additive manufacturing, advanced machining, and elimination of dissimilar metal welds (DMWs), EPRI, the US Department of Energy, and the UK-based Nuclear-Advanced Manufacturing Research Centre (Nuclear-AMRC) (together with a number of other industrial team members) will seek to demonstrate the hypothesis that critical sections of an SMR reactor can be manufactured/fabricated in a timeframe of less than 12 months and at an overall cost savings of >40% (versus today's technologies). Major components that will be fabricated from PM-HIP include: the lower reactor head, upper reactor head, steam plenum, steam plenum access covers, and upper transition shell.

The project aims to demonstrate and test the impact that each of these technologies would have on future production of SMRs, and explore the relevance of the technologies to the production of ALWRs, SMRs, GEN IV, Ultra-supercritical fossil, and supercritical CO2 plants. The project, if successful, may accelerate deployment of SMRs in both the USA and UK, and ultimately throughout the world for power production.

## Introduction

Over the past decade, EPRI, DOE, Nuclear-AMRC, and various OEMs and vendors have investigated a number of advanced technologies to support the manufacture of small modular reactors (SMRs). Advanced technologies including: electron beam welding for thick sections, powder metallurgy-HIP, diode laser cladding, dissimilar metal joining, and cryogenic machining are just a few of the examples technologies. Many of these technologies are now mature and can be readily demonstrated from production of SMRs.

In early 2016, EPRI and Nuclear-AMRC began assembly of a large project wherein the two organizations planned to demonstrate several of these advanced technologies aimed at the manufacture and fabrication of a 2/3rds scale SMR vessel. Their efforts were met with tremendous interest from industry and with co-sponsorship by the US Department of Energy (DOE) through its Advanced Methods for Manufacturing (AMM) program. The collaborative project began in earnest in early 2017 and is focused on developing/demonstrating three key advanced manufacturing technologies: electron beam welding for thick section components, powder metallurgy-HIP, and diode laser cladding, among a number of other manufacturing/fabrication technologies. The technologies are being demonstrated using NuScale Power's 50MWe (160MWth) SMR design.

The project aims to demonstrate and test the impact that each of these technologies would have on future production of SMRs, and explore the relevance of the technologies to the production of ALWRs, SMRs, GEN IV, Ultra-supercritical fossil, and supercritical CO2 plants. The project, if successful, may accelerate deployment of SMRs in both the USA and UK, and ultimately throughout the world for power production.

**Project Objectives**
Three key objectives were identified for this project. These include:
- Develop and demonstrate advanced manufacturing and fabrication technologies to rapidly accelerate the deployment of SMRs
- Develop/demonstrate new methods for manufacture/fabrication of a Reactor Pressure Vessel (RPV) which could lead to production of a vessel in under 12 months.
- Eliminate 40% of the costs of production of an SMR RPV, while reducing the overall schedule by up to 18 months.

These three objectives formed the basis formed for the entire project.

**Large Component Manufacture**
As described above, the NuScale Power reactor pressure vessel design, a 50MWe (160MWth), was selected for demonstration of various advanced manufacturing technologies at a 2/3rds scale. The NuScale Power RPV design was selected based upon its size and on the basis that it appears to be very near production. The RPV is shown in Figure 1 and consists of 8 major sections. In the current project, two assemblies, upper and lower (Figures 2 and 3), will be demonstrated. These two assemblies were selected based on the premise that the two assemblies would demonstrate many of the advanced manufacturing/fabrication technologies required for construction of an SMR and that assembly of the middle section would simply be redundant with regards to demonstrating these technologies.

Manufacture of the key components for the reactor vessel includes both conventional forging and PM-HIP. The breakdown of the key A508 low alloy steel components are as follows:

PM-HIP (A508, Grade 3, Class 1)
- Lower reactor head
- Upper reactor head
- Steam plenum
- Steam plenum access ports/covers

Hot Isostatic Pressing – HIP'17                                    Materials Research Forum LLC
Materials Research Proceedings **10** (2019) 224-234      doi: http://dx.doi.org/10.21741/9781644900031-29

- Upper transition shell (in four sections)

Forgings (SA508, Grade 3, Class 1)
- Lower transition shell
- Upper flange
- Lower flange
- PZR (pressurizer) shell

For the purpose of this paper, the discussion will only focus on manufacture of components using the PM-HIP process since forging is already a mature process. The forgings used in this project will be used for the upper and lower assemblies primarily for the purpose of demonstrating the joining and cladding technologies.

PM-HIP was considered for this project for a number of reason. The investigators believed PM-HIP provides a ready means to produce the upper and lower reactor heads in a near-net shaped condition minimizing hundreds of hours of machining from the manufacturing process, 2) it provides excellent properties often better than those that may be produced via forging, 3) the process can significantly reduce the time/schedule to produce a reactor head, and 4) the process provides excellent inspection characteristics. PM-HIP coupled with EB welding and heat treatment also provides an avenue to produce reactor components in sections, thereby eliminating the need for very large forgings such as those used in LWRs and ALWRs for the past 3 decades.

It should also be pointed out that PM-HIP is not currently accepted for production of A508 materials by ASTM or ASME Boiler and Pressure Vessel Code presently. One of the goals of the project is to develop sufficient mechanical, microstructural, and heat treatment data to support the development of an ASTM specification for A508 PM-HIP components and an ASME Code Case for A508 materials. Considerable work will be required to generate this supporting data and information. Preliminary research by the investigators has demonstrated excellent tensile and toughness properties can be readily achieved with the process however.

In 2017, two PM-HIP components are being manufactured: the upper head at 40% of full size and one-half of the lower head at 2/3rds size. The 50-inch (1.27m) in diameter upper head consists of 27 penetrations and weighs almost 2400lbs (1088kgs), is shown schematically in Figure 4. It is being manufactured at Synertech-PM at ~40% of its full size to demonstrate that it can be produced to near-net shape with today's HIP vessel technology. Only the size of the HIP vessel limits our ability to produce a larger component. Later in the project, the upper head will be produced in two one-half sections at a 2/3rds scale and EB welded together (more on this topic in the next section).

The lower head is also being produced via PM-HIP. At 2/3rds scale, the head is roughly 8ft (2.44m) in diameter and will be produced in two half sections and welded together as shown in Figure 5. The lower head upon completion will weigh approximately 4300lbs (1950kgs). Further discussion on EB welding of the head will be provided in the following section.

Other key 2/3rds-scale components that will be produced from PM-HIP include the steam plenum which will weigh almost 13,000lbs (5900kgs) and the upper transition shell which will weigh approximately 20,000lbs (9072kgs). The latter will be described in the following Fabrication section. The former will consist of five major sections that are produced via PM-HIP and welded together via EB welding (not covered in this paper however).

Additionally, the authors would like to point out that a full size steam plenum access port was produced under EPRI's Advanced Nuclear Technology (ANT) program in 2016. The 1200lb

(544kg) access port is shown in Figure 6 (still in the HIP capsule) and was produced from A508 powder. An access port will also be produced at 2/3rds scale for the upper assembly later in the project.

**Component Fabrication**
Two advanced fabrication technologies, electron beam welding (EBW) and diode laser cladding (DLC) are being developed/demonstrated in this project. EBW will be used to join half sections of the reactor heads, to perform major girth welds around the diameter of the vessel, and to join PM-HIP sections of the steam plenum and sections of the upper transition shell together. DLC, which will not be discussed herein, is being employed to apply the cladding to both the ID and OD surfaces of the reactor vessel.

EBW was selected as the joining process for this project due to three key attributes: 1) no filler wire is used with the process (eliminates the potential of weld embrittlement later in reactor service), 2) welds can be completed in a single pass, and 3) the investigators believe the technology can eliminate 80-90% of the welding time necessary for completion of the reactor vessel assembly.

As mentioned earlier, the lower and upper heads for the 2/3rds scale assembly will be produced using one-half sections (produced by PM-HIP) and welded together via EBW. Figure 5 provides a pictorial view of the lower head, while Figure 7 provides a view of the upper head. Each weld will be performed in the horizontal welding position as shown in Figure 8. Following welding, the assembly will be annealed, quenched, normalized and then tempered to produce final properties and to remove any evidence of the weld.

Assembly of the entire SMR vessel would require 10 major girth welds. Since the middle RPV assembly is not being considered in this project, only 5 girth welds will be demonstrated. The first girth weld will be performed to attach the lower flange to the lower flange shell (Figure 9). This is performed in the upside down position. Next, the lower reactor head will be attached to the lower flange as shown in Figure 10 (again shown in the upside down position). This completes one-half of the lower reactor assembly.

The upper one-half of the lower reactor assembly is comprised of an upper transition shell and a large flange. The upper transition shell consists of four individual sections which will be joined using vertical EB welding (Figure 11). Next, the upper transition shell will be welded to the upper flange once again using a horizontal girth weld (Figure 12). This completes the second-half of the lower assembly. At this point, both halves of the lower assembly will be heat treated, final machining will be performed, and the two flanges are ready to be bolted together. This will complete the first phase of the project.

The upper assembly, which is the more difficult of the two assemblies, will be produced and assembled in Phase 2 of the project which is slated to begin in late 2018. For the purpose of this discussion, individual girth welds will be applied to join the steam plenum to the PZR shell and the PZR shell to the upper head. There are a number of other EB welds which will be performed to assemble the steam plenum, but they will not be covered in this paper. Approximately 20 EB welds will be performed on thick section components ranging from 75-110mm in this project

**Project Status**
Much of 2017 has focused on detailed planning of various work packages to support the project in the areas of machining, EBW, DLC, NDE, PM-HIP, etc. The work packages have been completed and development of mockups/test pieces is currently in progress to support each of these technologies. Additionally, forgings for the project have also been ordered from Sheffield Forgemasters and are anticipated during Q1-2018.

For the PM-HIP portion of this project, the focus has been on production of a one-half section of the lower head and production of the 40% upper head section. Both of these components will provide valuable insight into production of full sections of these components to support the overall project moving forward.

**Project Summary**

Several advanced manufacturing and fabrication technologies have been covered in the subject Advanced Manufacturing and Fabrication project. Two key technologies: PM-HIP and electron beam welding (EBW) have been covered in this paper. Both technologies have the potential to completely change the way industry fabricates reactor vessels today, to rapidly accelerate deployment of SMRs, and to significantly reduced costs and time toward production of a SMR. PM-HIP is being used to manufacture/demonstrate five major sections of the reactor vessel including:

- Lower reactor head
- Upper reactor head
- Steam plenum (in five sections)
- Steam plenum access ports/covers
- Upper transition shell (in four sections)

It is believed that PM-HIP can offer dramatic improvements in both properties and in time-delivery schedule for reactor parts.

EBW is being employed to join reactor head sections, produce vertical welds to join shell sections, and to perform major girth welds to join sections of the reactor. The EBW technology is being considered in this project for the following reasons: 1) it eliminates the need for welding filler material, 2) it can be performed in a single weld pass, and 3) it has the potential to eliminate 80-90% of the welding time required for reactor vessel fabrication.

Together these two advanced manufacturing/fabrication technologies have the potential of making SMRs competitive with other lower cost energy production methods such as natural gas.

**Acknowledgements**

The authors would like to recognize Tom Miller, Alison Hahn, and Tim Beville from the U.S. Department of Energy for their support and efforts in this project.

Additionally, the authors would also like to recognize Vern Pence, Matt Mallet, and Tamas Liskai from NuScale Power for their input, advice, and engineering input provided to the project.

*Acknowledgment: "This material is based upon work partially supported by the Department of Energy under Award Number DE-NE0008629."*

*Disclaimer: "This report was prepared as an account of work partially sponsored by an agency of the United States Government. Neither the United States Government nor any agency thereof, nor any of their employees, makes any warranty, express or implied, or assumes any legal liability or responsibility for the accuracy, completeness, or usefulness of any information, apparatus, product, or process disclosed, or represents that its use would not infringe privately owned rights. Reference herein to any specific commercial product, process, or service by trade name, trademark, manufacturer, or otherwise does not necessarily constitute or imply its endorsement, recommendation, or favoring by the United States Government or any agency thereof. The views and opinions of authors expressed herein do not necessarily state or reflect those of the United States Government or any agency thereof."*

Hot Isostatic Pressing – HIP'17                                    Materials Research Forum LLC
Materials Research Proceedings **10** (2019) 224-234        doi: http://dx.doi.org/10.21741/9781644900031-29

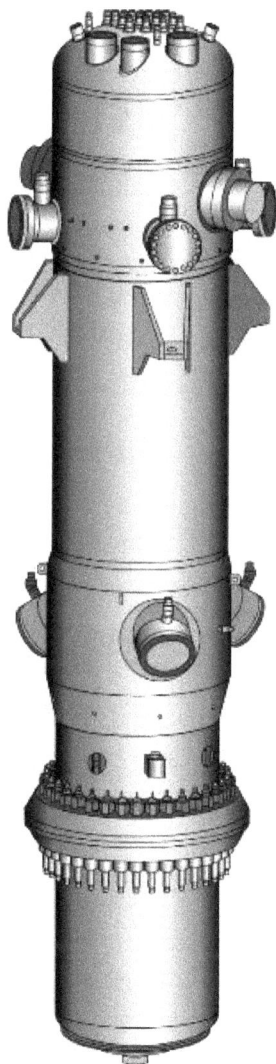

*Figure 1. NuScale Power Reactor Pressure Vessel (Courtesy of NuScale Power).*

Hot Isostatic Pressing – HIP'17                                    Materials Research Forum LLC
Materials Research Proceedings **10** (2019) 224-234       doi: http://dx.doi.org/10.21741/9781644900031-29

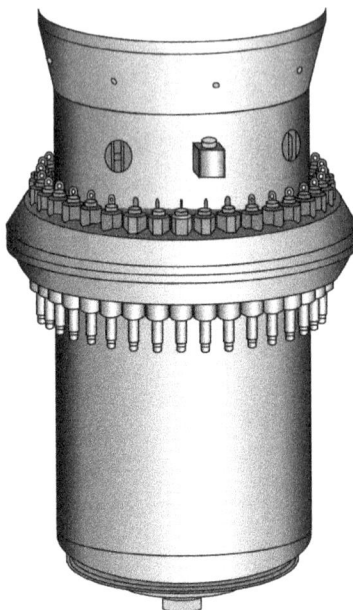

*Figure 2. Lower Reactor Assembly (Courtesy of NuScale Power).*

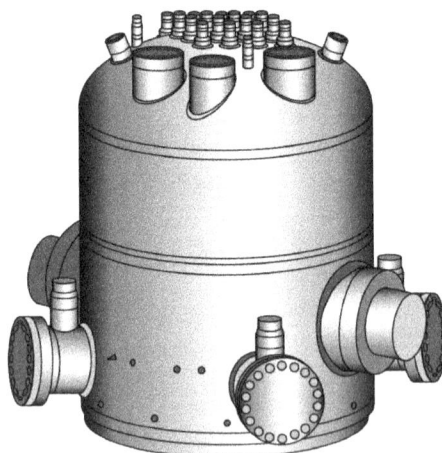

*Figure 3. Upper Reactor Assembly (Courtesy of NuScale Power).*

Figure 4.  Schematic of the upper reactor head at 40% size

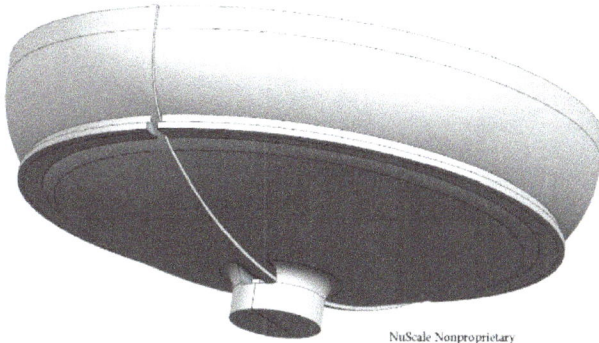

NuScale Nonproprietary
© 2017 NuScale Power, LLC

*Figure 5.  Lower reactor head produced in 2 halves.*

*Figure 6. A508 SMR Steam Access Port produced via PM-HIP.*

231

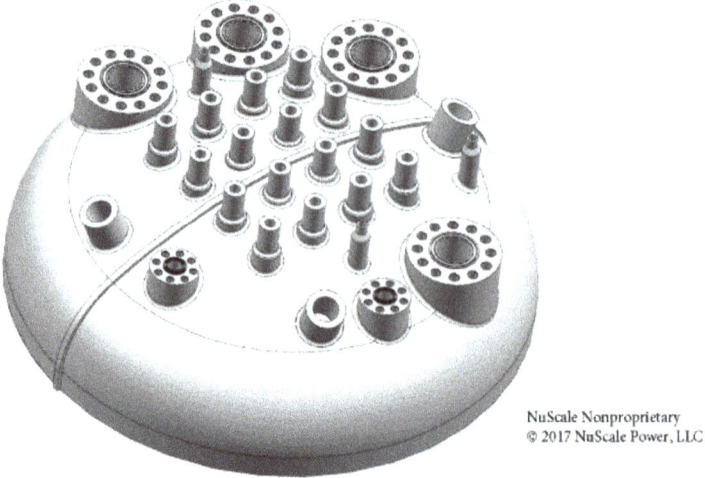

NuScale Nonproprietary
© 2017 NuScale Power, LLC

*Figure 7.  Schematic of the upper reactor head produced in 2 halves.*

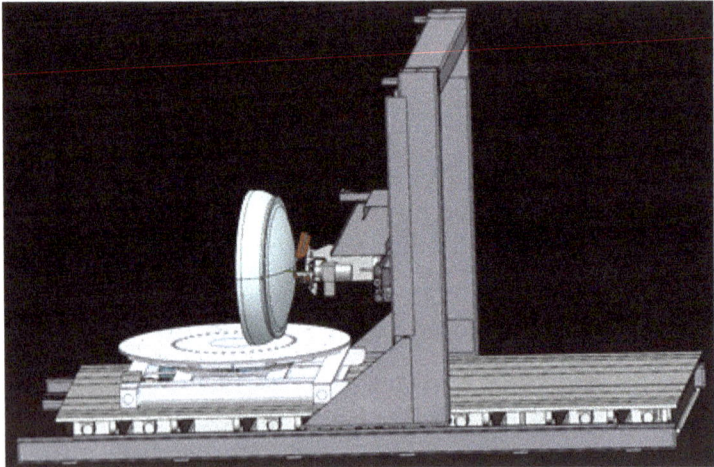

*Figure 8.  EB welding of the lower and upper heads will be performed in the horizontal position.*

Hot Isostatic Pressing – HIP'17                                     Materials Research Forum LLC
Materials Research Proceedings **10** (2019) 224-234        doi: http://dx.doi.org/10.21741/9781644900031-29

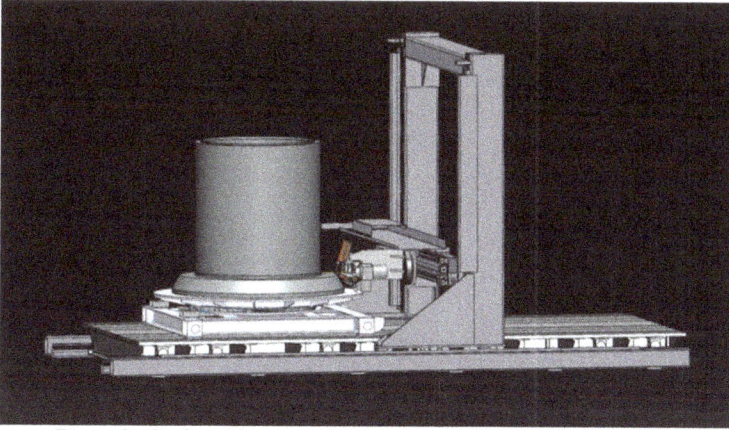

*Figure 9. Girth weld joining the lower flange to the lower flange shell*

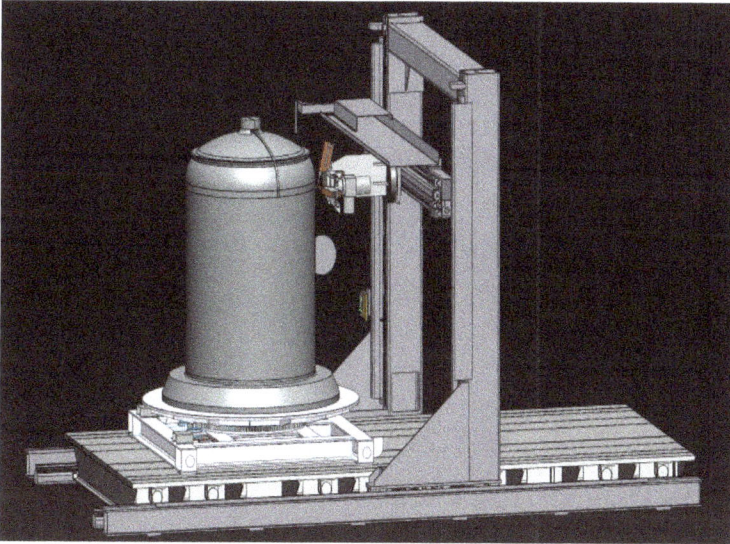

*Figure 10. EB welding will also be used to join the lower head to the lower flange shell.*

*Figure 11. Upper transition shell which consists of 4 individual PM-HIP sections joined via vertical EB welds. EB welds are shown in blue.*

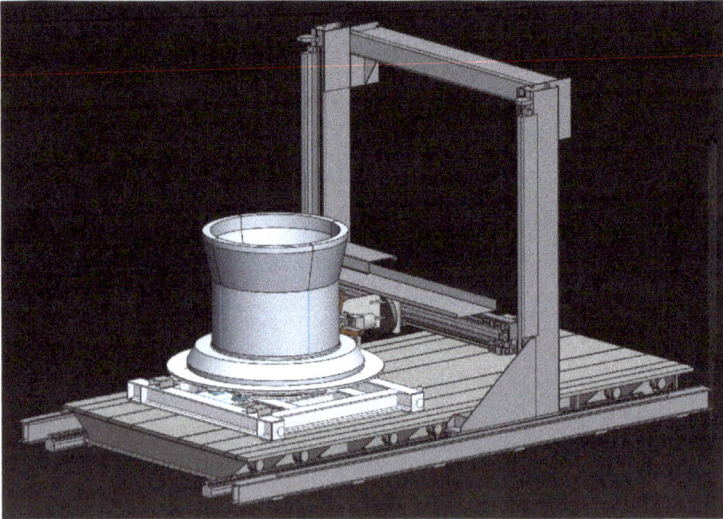

*Figure 12. Attachment of the upper transition shell to the upper flange will be completed using the EB process.*

Hot Isostatic Pressing – HIP'17                                              Materials Research Forum LLC
Materials Research Proceedings **10** (2019) 235-241          doi: http://dx.doi.org/10.21741/9781644900031-30

# Mechanical Strength Evaluation of Superconducting Magnet Structures by HIP Bonding Method

Kaoru Ueno[1,a*], Tsuyoshi Uno[1,b], Katsutoshi Takano[2,c] and Norikiyo Koizumi[2,d]

[1]Metal Technology Co. Ltd. Ibaraki Plant,
276-21 Moto-Ishikawa-cho, Mito-city, Ibaraki pref. 310-0843, Japan

[2]National Institute for Quantum and Radiological Science and Technology,
801-1 Mukoyama Naka-city, Ibaraki pref. 311-0193, Japan

[a]kueno@kinzoku.co.jp, [b]tuno@kinzoku.co.jp, [c]takano.katsutoshi@qst.go.jp,
[d]koizumi.norikiyo@qst.go.jp

**Keywords:** HIP, Diffusion Bonding, FM316LNH, ITER, Radial Plate

**Abstract.** MTC proposed that production of the International Thermonuclear Experimental Reactor (ITER) toroidal magnetic field coils support structure "Radial Plate (RP)" using the HIP diffusion bonding. HIP allows production cost reductions in materials and machining time. However, insufficient strength of the HIP diffusion bonded sections was an issue. In this study, changes in bonding strength and base material strength were investigated using HIP treatment temperatures as parameters. Mechanical property tests and microstructure observation of the cross sections at the bonded areas after HIP treatment were conducted and as a result it was found that the bonding strength equivalent to the mechanical strength of the base metal can be obtained at 1300[°C]. However, the yield strength at room temperature of the base material after HIP treatment decreased by 24%. Since crystal grain coarsening occurs due to heat input by HIP, design considerations are required to guarantee the specified strength. Furthermore, a full-size mock-up of RP segments was fabricated by HIP. It was confirmed that the yield strength and fracture toughness at 4 [K] at the bonded sections of the mock-up satisfied the specified values required for the RP and uniform quality was obtained. From this study, it was proven that HIP diffusion bonding can be applied to the fabrication of RP segments.

## 1. Introduction

The Radial Plate (RP) is a structure for accommodating and supporting the conductors of the toroidal field coils of the nuclear fusion reactor at ITER. Grooves for accommodating conductors are provided in a D-shaped structure with a width of 9 [m], a height of 13 [m], and a thickness of 100 [mm] (Figure 1.1). The D-shaped structure is fabricated by welding the RP segments together. As RP are used within a strong electromagnetic field and at cryogenic temperatures, they must have a high yield and tensile strength. For this reason, FM316LNH stainless steel (Table 1.1) was chosen.

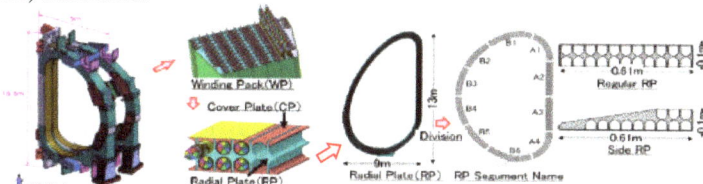

*Figure1.1  ITER TF Coil Radial Plate and Cover Plate [1]*

Hot Isostatic Pressing – HIP'17                                    Materials Research Forum LLC
Materials Research Proceedings **10** (2019) 235-241        doi: http://dx.doi.org/10.21741/9781644900031-30

The material cost of FM316LNH is high because it requires high purity raw materials and advanced forging technology. Moreover, since the RP has a curved shape with U-shaped grooves, machining from a forged plate has a material yield of about 20% [2]. Therefore, Masuo and Takahashi [3,4] proposed streamlining production by reducing material and machining costs, using HIP diffusion bonding to form materials to a near-net-shape. However, insufficient strength of the HIP diffusion bonded sections was an issue. In this study, trial pieces and a mock-up were made aiming at improving the strength of the bonded sections. In this presentation, we will report on the following two points:

• Investigation of bonding conditions with processing temperatures as parameters in HIP diffusion bonding between two blocks of FM 316LNH; and
• Results of RP segment mock-up production using HIP diffusion bonding.

*Table 1.1  Chemical compositions of FM316LNH [5]*

| material | C | Si | Mn | P | S | Ni | Cr | Mo | N | C+N |
|---|---|---|---|---|---|---|---|---|---|---|
| mass % | 0.030 max. | 2.00 max. | 2.0 max. | 0.035 max. | 0.020 max. | 10.00 -14.00 | 16.00 -18.50 | 2.0 -3.0 | 0.15 -0.20 | 0.180 Min |

*Table 1.2  Mechanical properties of FM316LNH (Hot Rolled Plate, Thickness<200mm) [5]*

| | Yield Strength(MPa) | | | Charpy Absorbed Energy(J) | Fracture Toughness $K_{IC}$ (MPa√m) |
|---|---|---|---|---|---|
| Test temperature | RT | 77K | 4K | 77K | 4K |
| Design strength | 280 | 705 | 900 | - | - |
| Reference Design Value | - | - | - | 272 | 255 |

## 2. Investigation of bonding temperature

Processing conditions were sought with the goal that the base material sections and the bonded sections satisfy the design strengths (Table 1.2) for FM316LNH.

**2.1 HIP treatment.** Test pieces were prepared by changing the HIP bonding temperature from 1050 [°C] to 1300 [°C] in increments of 50 [°C]. The same HIP device was used throughout.
HIP processing conditions:  X [°C] × 118 [MPa] × 4 [hrs.]X = 1050, 1100, 1150, 1200, 1250, 1300.

**2.2 Material.** FM316LNH hot-rolled plate was used for the test pieces. Table 2.1 shows the chemical composition of the material. The bonded material test piece (B) configuration is shown in Figure 2.1. A block of the base material (M), 220 [mm] × 115 [mm] × 20 [mm] was prepared and HIP treatment was performed on the base material together with the bonded material at the same time. Tensile test, fracture toughness test, Charpy impact test, crystal grain size measurement tests, and cross-section observation of bonded sections were performed on the base material section (M) and the bonded section (B) of each test piece and the states for the respective bonding temperature were compared (Table 2.2). Particularly in the fracture toughness test and the Charpy impact test, the bond is expected to be weak because the grain boundaries are aligned along the bonded boundary. Therefore, in order to confirm the state in which the crystal grains grow beyond the bonding interface and the boundary disappears, the cross sections of the bonded sections were observed. In order to compare these values before and after the HIP treatment, the same test was performed on the material before and after the HIP treatment.

Hot Isostatic Pressing – HIP'17                                    Materials Research Forum LLC
Materials Research Proceedings **10** (2019) 235-241        doi: http://dx.doi.org/10.21741/9781644900031-30

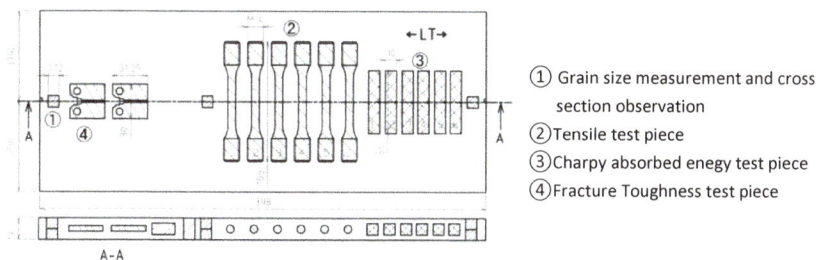

① Grain size measurement and cross
   section observation
② Tensile test piece
③ Charpy absorbed enegy test piece
④ Fracture Toughness test piece

*Figure 2.1  Bonded Material Test Piece Configuration  (B)*

*Table 2.1 Chemical compositions of material*

| material | C | Si | Mn | P | S | Ni | Cr | Mo | N | C+N |
|---|---|---|---|---|---|---|---|---|---|---|
| mass% | 0.024 | 0.54 | 1.53 | 0.018 | 0.0003 | 12.95 | 16.48 | 2.54 | 0.172 | 0.196 |

*Table 2.2 Bonded material and Base Material Test*

| Test Item | Condition | Section | n | Method |
|---|---|---|---|---|
| Grain Size | Arbitrary | M | 1 | JIS G 0551 |
| Bonding Surface | Arbitrary | B | 6 | Micro graphic |
| Yield Strength | RT, 77K, 4K | M,B | 2 | JIS Z 2201 |
| Charpy Absorbed Energy | RT, 77K | M,B | 3 | ASTM E 23 |
| Fracture Toughness $K_{IC}$ | 4K | M,B | 2 | ASTM E 1820 |

## 2.3 Result
### 2.3.1 Grain Size and Yield Strength

Figure 2.2 shows the grain size with respect to HIP bonding temperatures and the changes in the yield strength (Y.S.) average at room temperature (RT) and 4 [K].

The Y.S. of the bonded section was the same strength as that of the base material section under all conditions. However, when compared with the material before HIP treatment, Y.S at RT decreased. Since grain boundaries impede dislocation movement, Y.S. tends to increase as the crystal grain size becomes smaller. (J.N. Petch Equation [2]) It is thought that the phenomena attributable to this occurred in the base material section. For the test pieces subjected to HIP treatment at 1300 °C, the average value of the Y.S. decreased by 24% at RT, compared with the material before HIP treatment, whereas

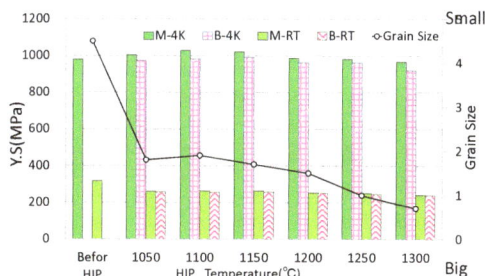

*Figure 2.2 Changes in grain size & Y.S.(RT,4K)*

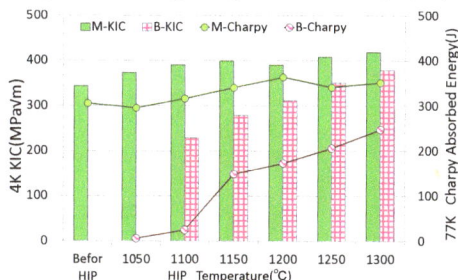

*Figure 2.3 Changes $K_{IC}$(4K) and Charpy absorbed enrgy (77K)*

the average value of Y.S. decreased by 10% at 77 [K], and merely 1% at 4 [K]. As the test temperature decreases, it is presumed that the influence of the impediment to dislocation movement by grain boundaries becomes smaller and the decrease rate of Y.S. against the value of Y.S before HIP treatment decreases.

### 2.3.2 Fracture Toughness Value and Charpy Absorbed Energy

Figure 2.3, shows the average of the fracture toughness value at 4 [K] and Charpy absorbed energy at 77 [K] at the bonded section.

The bonded sections fracture toughness and Charpy absorbed energy increase with the rising HIP bonding temperature.

The fracture toughness values at the bonded section satisfied the design reference values at a bonding temperature of 1150 °C and above. On the other hand, Charpy absorbed energy values at the bonded section didn't meet the design reference values. Both the fracture toughness values and the Charpy absorbed energy values of the base material section satisfied the design reference values under all bonding conditions. It is noted that the strength of the base material section tends to increase with HIP bonding temperature and this is presumed to be caused by the coarsening crystal grains. The influence of crystal grain size on fracture toughness values and Charpy absorbed energy values is currently under investigation. The strength improvement due to the rise in the bonding temperature was particularly conspicuous at the bonded section. This is thought to be due to the fact that improvement of the bonding interface affects the strength.

### 2.3.3 Crystal Grain Size and Condition of Bonded Interface

The relationship between crystal grain sizes and HIP bonding temperature is shown in Figure 2.2. Table 2.3 shows the micro structure of the bonded section in cross-section and the fracture surface of the Charpy test piece for each HIP bonding temperature. Along with the coarsening of crystal grains accompanying the rise in bonding, the crystal grain progresses and the bonding interface disappears. From the fracture surface of the impact test piece at the bonded section, it can be seen that the fracture of the bonded section changed to have a ductile fracture behavior as the bonded interface disappeared. From this, it is believed that the state of the bonding interface is related to the bonding strength, and it is possible to judge whether the bonding state is good or bad by observing the state of the bonding interface.

*Table 2.3 Bonding interface state & Charpy fracture surface in each temperature range*

| HIP Temperature | 1100°C | 1250°C | 1300°C |
|---|---|---|---|
| Bonding Surface | | | |
| Charpy Fracture Surface | | | |

**2.3.4 Summary.** With the rise in HIP temperature, the following phenomena occurred:
- Coarsening (growth) of crystal grains,
- Disappearance of the bonded interface,
- Reduction in Y.S. at RT, 77 [K],

- Improvement of fracture toughness and Charpy absorbed energy values of the base material and bonded sections.

As the crystal grains coarsen and the bonded interface disappears, the toughness of the bonded section is improved. In view of this result, it is thought that the bonding state can be judged by confirming the crystal state near the bonding interface.

## 2.4 Optimization of Bonding Temperature

- Based on the result in this section, the best HIP bonding temperature is 1300 [°C]. It provides the highest strength at 4 [K],
- When HIP diffusion bonding is applied to this material, it is necessary to design the strength whilst considering reduction of Y.S. due to heat input.

## 3. Manufacturing of full-size mock-up

Based on the results in Section 2, in order to confirm that consistent results can be obtained in a full-size structure, RP segment mock-up was fabricated by HIP diffusion bonding.

**3.1 Mock-up Structure.** The capsule structure of the mock-up and its fabrication flow are shown in Figure 3.1. The material was taken from the same lot as in Section 2. After HIP treatment, machining was carried out and the required accuracy was confirmed.

**3.2 HIP Processing Condition.** HIP process condition: 1300 [°C] × 118 [MPa] × 4 [hrs.]

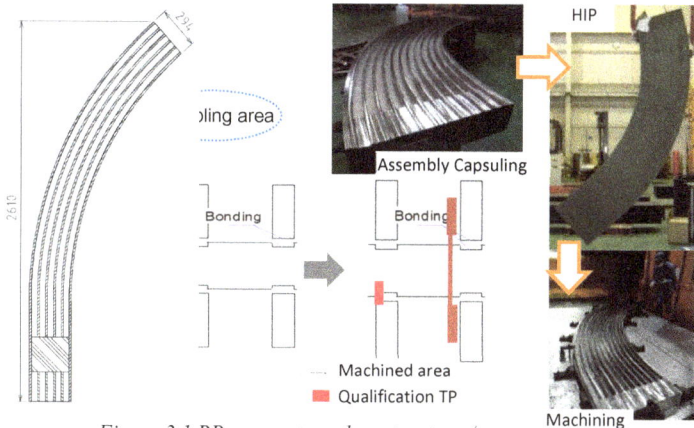

*Figure 3.1 RP segment mockup structure / production flow*

**3.3 Quality Confirmation Test.** Test pieces were taken from the mock-up after HIP treatment and quality confirmation tests on the bonded sections (Table 3.1) were performed. The areas removed are shown in Figure 3.1. From Table 3.2, it can be confirmed that the growth of crystal grains progresses and the bonded interface disappears. As shown in Table 3.3, Y.S. at 4 [K] and fracture toughness values satisfied the specified values of the material in all test pieces. On the

Hot Isostatic Pressing – HIP'17                                    Materials Research Forum LLC
Materials Research Proceedings **10** (2019) 235-241        doi: http://dx.doi.org/10.21741/9781644900031-30

other hand, Y.S. at RT and 77 [K] were lower than the specified values. However, there is little variation in the strength, and no substantial difference is observed, compared with the strength of the base material section and the bonded section of small-size bonded bodies (Section 2). In addition, all the tensile test pieces were fractured in the base material sections with basically the same strength as that of the base material parts. As a result, it was confirmed that uniform bonding quality was obtained through the entire structure.

*Table 3.1 Quality Confirmation Test*

| Test Detail | n | Test Detail | condition | n |
|---|---|---|---|---|
| Composition Analysis | 1* | Yield strength | RT, 77K ,4K | 6/area |
| Grain Size | 1* | Fracture Toughness K$_{IC}$ | 4K | 8/area |
| Bonded Surface | 26/area | **One segment in the entire mock-up | | |

*Table 3.2 Bonding interface of mock-up*

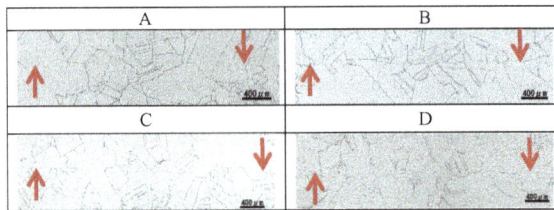

*Table 3.3 Mock-up bonded part mechanical strength results (average value)*

| Sample | Y.S.(MPa) | | | Fracture Toughness K$_{IC}$ 4K (MPavm) |
|---|---|---|---|---|
| | RT | 77K | 4K | |
| Specified value of material* | >280 | >705 | >900 | >180 |
| Material before HIP treatment** | 318.1 | 745.2 | 975.7 | 343.0 |
| 1300°C HIP Base material part ** | 243.3 | 673.1 | 967.4 | 418.9 |
| 1300°C HIP Bonded part ** | 241.4 | 659.9 | 919.6 | 378.7 |
| Area A | 238.2 | 644.8 | 931.8 | 288.5 |
| Area B | 237.3 | 646.8 | 932.3 | 325.6 |
| Area C | 235.6 | 644.0 | 932.2 | 331.5 |
| Area D | 232.5 | 634.5 | 918.8 | 334.1 |

*Reprinted from Table 1.2    **Reprinted from Section 2

**3.4 Reduction Factor.** Table 3.4 shows the allowable stress reduction factor estimated from the yield strength of the full-size mock-up compared with pre-HIP material. The reliability of the factor is greater than or equal to 99% by the point of μ-2.33σ in the probability function. Here, μ is the arithmetic mean and σ is the corrected sample standard deviation.

*Table 3.4 Allowable Stress Reduction Factor*

| Temperature | RT | 77K | 4K |
|---|---|---|---|
| Average value | 0.74 | 0.86 | 0.95 |
| Standard Deviation | 0.01 | 0.01 | 0.01 |
| 2.33σ | 0.02 | 0.02 | 0.02 |
| Reduction Factor | 0.72 | 0.84 | 0.92 |

The allowable stress reduction factor is given by; $1.955 \times 10^{-6} T^2$ - $1.295 \times 10^{-3} T + 0.9311$. Where, $T$ expresses temperature [K]. The allowable stress of HIP bonded structure is given by multiplying the allowable stress reduction factor onto the allowable stress of the structure evaluated before the HIP treatment.

### 3.5 Summary: Production of full-size Mock-up
- Mock-up Fabrication: HIP: 1300 [°C] × 118 [MPa] × 4 [hrs.]
- Constant quality can be obtained regardless of product size.
- Y.S., fracture toughness satisfies standard values at 4 [K].
- Y.S. at RT and 77 [K] fell below the specified value of the material. .
- The allowable stress of HIP diffusion bonded structure can corrected by the allowable stress reduction factor.

## 4. Conclusion
In order to streamline the production of RP segments, we examined the HIP temperatures to optimizing the HIP processing conditions in diffusion bonding of FM316LNH materials together. The following conclusions were drawn from comparison of various mechanical strengths obtained with small test pieces for each temperature parameter. Fabrication was verified by a mock-up structure.
- HIP diffusion bonding temperature for the highest mechanical property at 4 [K] is 1300[°C].
- Despite the size and structure of the HIPed parts uniformed bonding was achieved.
- Bonded strength depends on grain development at the bonding interface; the bonded state can be determined by observing the bonded interface.
- Coarsening of crystal grains occurs during HIP treatment and the yield strength of the base material part tends to decrease with temperature increases.
- In the design of the HIP diffusion bonded structure, it is necessary to consider reduction of the yield strength especially at room temperature and 77K, due to the crystal grains coarsening during HIP treatment. In the design of the HIP diffusion bonded structure, the allowable stress can be obtained by multiplying the limit value of the stress intensity with the allowable stress reduction factor.
- Future work will be to collect fatigue strength data at operating conditions.

### References
[1] K. Takano, et al. "Development of Manufacturing Technology of Radial Plate in Superconducting Coil for Fusion Reactor by Diffusion Bonding by Hot Isostatic Pressing (HIP)" YOSETSU (2012)

[2] J.N. Petch. Journal of the Iron and Steel Institute, Vol.2 (1953)

[3] H. Masuo, et. al. "Application of HIP diffusion bonding technique in the development of nuclear fusion reactor" HIP'11(2011)

[4] M. Takahashi. et al. "Fabrication of super electromagnetic coil support using HIP diffusion bonding" HIP'14(2014)

[5] The Japan Society of Mechanical Engineers Standard (2008)

# Keyword Index

# Author Index

# About the Editors

### Gerry Triani

Gerry is the Technical Director of ANSTO Synroc, an Australian technology that utilises HIP processing for the treatment of nuclear waste. As a materials scientist with 30 years of experience, his research has been focussed on the structure-function relationship of materials and their application in nuclear, medical and energy sectors.

### Pranesh Dayal

Pranesh is a professional materials engineer with over 15 years of experience in the field of steel industry, advanced materials synthesis, materials characterization and nuclear industry. Currently he a member of ANSTO Synroc wasteform design team.

www.ingramcontent.com/pod-product-compliance
Lightning Source LLC
Chambersburg PA
CBHW060357220326
41598CB00023B/2952